Amazing UFO Sightings in the Hudson Valley Hidden technology & the coming great deception

By Steve Marconi

Order this book online at www.trafford.com
or email orders@trafford.com

Most Trafford titles are also available at major online book retailers.

Print information available on the last page.

ISBN: 978-1-4251-6406-5 (sc)

Trafford rev. 04/13/2021

www.trafford.com
North America & international
toll-free: 844-688-6899 (USA & Canada)
fax: 812 355 4082

Amazing UFO sightings in the Hudson Valley Hidden technology & the coming great deception

One man's detailed sightings and shocking research opinions with two close encounters, and multiple witness sightings.

The incredible UFO flap of 1988 and continued sightings with new revelations about the 1975 UFO landing in North Hudson Park, New Jersey.

Includes powerful research opinions on the Bible, ancient UFOs, and scientific conspiracy.

By Steve Marconi

Preface

From the Hudson River opposite midtown Manhattan, to a quiet upstate New York country community, shocking UFO sightings are described in exact detail. Multiple eyewitnesses saw apparent antigravity craft operating in an area near military bases. Unmarked black helicopters flew before and after certain sightings on the same flight path, at the same altitude, in the same direction.

On more than one occasion a military transport aircraft apparently escorts one of these craft flying behind it, or more accurately, floating and rocking behind it. A technology producing invisibility is demonstrated and witnessed by more than one. Other strange occurrences baffle residents in the area, including a low pitched hum in the morning hours, and an explosion during a snowstorm sending out a blinding white light in all directions.

Military aircraft performed hoaxes simulating the UFOs, flying with unusual light formations on them. In other instances airplanes flew in formation attempting to confuse the viewing public. In these cases the sky watchers were not fooled, they clearly discerned the difference between conventional and unconventional craft. A small group of friends and neighbors who were skilled observers watched the skies and were not fooled. Such is the case on a local basis throughout the world. Sometimes the public figures out what's going on much better than the authorities would hope, but the media continues its whitewashed story on a national basis, and on a worldwide basis.

A skilled observer and eyewitness to 14 sightings was traumatized by what he witnessed and went on a path of research throughout his sightings which led him into amazing scientific revelations of suppressed technology, shocking information on past inventors, and amazing insights from ancient texts and Biblical scripture. His research reveals hidden technology and produces powerful opinions relating to Bible prophecy and end times.

He also discovered an amazing revelation about a UFO landing in the area where he was raised as a boy. This phenomenon has followed him from boyhood into adulthood, then he turned around and shown the spotlight of research and the spirit of truth on the things he saw, and came up with some really powerful insights. This book offers a unique perspective and powerful food for thought. A truth seeker lets loose with his story and some very detailed sightings, important statements, research findings, and opinions. You won't be able to put this book down until you've finished it.

This book is the frank and up front testimony of a multiple UFO eyewitness, expressed in extreme detail.

This book is written with Coast to Coast AM listeners in mind, a radio show the author is grateful to, for airing the testimonies of people who have seen the paranormal world for themselves, and share it with all. Decades of observing the sky have turned the author into a skilled observer. Couple this with a precision memory and he gives amazing details of some hair-raising sightings including two close encounters shocking enough to induce cardiac arrest.

Very clear sightings of very unconventional craft are given. Many occurrences in between sightings are detailed, including black helicopters without markings, military jets in formation, military aircraft accompanying the UFO, and other strange occurrences such as the hum and the sky quake. The side bar information in between UFO sightings shows what a busy time period this was in the area, especially in 1988. The area where the author lived was very close to two military installations.

The author discusses ancient technology in relation to the Coral Castle in Florida. His research reveals technological secrets about UFOs. He then gives amazing revelations about the UFO landing in North Hudson Park, New Jersey, in 1975 that have never been revealed until now. He then gives his opinions about ancient scriptures and the Bible in relation to UFOs. A unique viewpoint is given from a biblical standpoint concerning UFOs that will be of great interest to those who have studied the Bible in depth, and to those who have faith in it in general. Scientific suppression of inventions by global elites is discussed. Fiery opinions are given in strong terms without mincing words. These opinions are very thought provoking and have not been previously given in any UFO book. The author has read many UFO books and has seen most every UFO documentary made, and has kept track of UFO reports in the media for 20 years. Showing a vast and rare knowledge of advanced science and technology, and having knowledge of great inventors of the past; the author gives insight into UFO technology.

Leaning towards the Earth based origin of UFOs the author touches on secret societies and suppressed technology, and builds up towards a biblical apocalypse of judgment upon a grand deception in the final war between good and evil. We can choose evil controllers, or God's Garden of Eden. The machines of the false gods or the paradise of the real God, the choice will soon be yours in the coming great deception. The war between the Gods will be repeated just like at the time of the 10 plagues of Egypt. Will you follow the gods of technology, or the God of prophecy? Will you choose the gods of machines, or the God of nature? The showdown is coming! An amazing unique book unlike any UFO book you've read. Traumatized by incredible sightings and close encounters, the author was forced to grapple with these questions and you will be also.

Everyone who has faith in the Bible and has seen a UFO must read this book.

Those who don't know the Bible and have seen a UFO must hear this perspective.

Those who want insight into UFO technology must read this book!

This book is factual. The sighting descriptions are true. The research and opinions are shocking. Here is a different kind of UFO book. It comes from a thinking, individual, private, eye- witness researcher, who has studied suppressed technologies, global elites and conspiracy, and ancient scripture.

It comes from the public citizenry rather than from government or military sources. Rather than hearing the same old lines we've been given for years, you will hear a perspective no one in the official world wants to discuss. Sometimes better information comes from an individual, without affiliations to official organizations. You will be given food for thought on subjects that have been avoided, on purpose, by those in power and those who control a lot of the research of this subject.

After 30 years of being told that we're just seeing Venus and "swamp gas" or hallucinating, we were then told that aliens are here and the media's been pushing aliens on us for the subsequent 30 years. This author doesn't buy any of the official lines coming from the media or the information controllers, or the ideas coming from many other researchers. The search for the truth is an individual one, and in this book you will be steered in directions others are afraid to go. After being given extremely detailed sightings of craft which were obviously antigravity machines, and could not be mistaken for otherwise by the eyewitnesses, you then learn about suppressed technology from brilliant inventors of the past. You will hear about some of the aspects of the UFO phenomenon that have been buried deep under official lines of controlled information. Hidden technological information is brought to the surface, which should have given us free energy and antigravity long ago.

A private citizen reading about brilliant inventors of the past shares his knowledge after 18 years of researching UFOs during and after many incredible sightings, including two amazing close encounters. His sightings were the catalyst for his research, because after seeing, a curious person wants to know what's going on. After giving his opinion on the Coral Castle in Florida, he gives you his revelation about a UFO landing in 1975 which no one else has ever mentioned or become aware of until now.

7

The last part of the book deals with powerful opinions on the subject coming from a solid knowledge of the Bible and supporting ancient scriptures. Contrary to modern spin, science and ancient scripture lineup perfectly, and ancient prophecies accurately predicted the future in this hyper technological age. A natural born scientist who can link up past and present, and see into the future, brings you on a journey of knowledge you will never forget. The past repeats itself like a huge cycle only this time the technology is greater, so the judgment and cleansing will be greater. This cycle has happened before with Atlantis and Lemuria, and with ancient Egypt. There is a final war between good and evil and a final judgment. Those who interfered with the peace of Earth, and have polluted it, will be cut off in the final judgment. You will never look at the subject of UFO's the same after reading this book.

Technology of this type is not our savior; it is the tool of our enemies the way it is being used. Our real savior is a God of love who gave life for free, and will perfect it for all of us in a clean ecology of sharing and love after our enemies are gone, and then we will enjoy eternal peace. A positive faith for the future and science meet here. The Native American and Biblical perspectives concerning the final cleansing of Earth meet here.

Justice is coming after the final technological war. The war in our atmosphere has begun and it reaches into space, where the enemies of God and mankind seek to hide, but they will be dealt with no matter where they hide in the final judgment and cleansing of the Earth. Join a unique researcher on this journey and see past the spin we've been told, and the suppression of incredible, clean, electrical anti-gravity technology, and limitless free energy we have wrongly endured!

CONTENTS

My first sighting

1983

I was with my friend Alan C. I believe it was Sunday June 9th, on a clear and sunny day, there was no mistaking what we saw. We were in Cliffside Park, New Jersey, off of Pine Street; on the cliff at a place they call the slanted rock. We were playing guitar and singing songs when all of a sudden we caught sight of an object flying over the Hudson River about halfway over the river.

This object was unlike anything we had ever seen, it was a black cylinder, and the shape was like a cigarette. It was about 75 ft. long and maybe 15 ft. wide. It was rotating and glinting in the sun as though it were metallic; it was a flat black color and looked like a gloss black at the edges. It was also tumbling end over end straight across the sky in a circular motion. It moved slowly, perhaps 30 mi. per hour. The object made no sound whatsoever. We looked at it in amazement asking each other, 'What the heck is that thing'? We couldn't believe what we were seeing.

When it got to our side of the river, it went over the houses only about an eighth of a mile away from us. It was flying only about a hundred feet above the houses. At this point two more friends came by,

Mark M. and Larry C. They saw the object for about 30 seconds, but Alan and I had watched it for two full minutes. The object continued flying over the area of Nungessers, where Cliffside Park meets North Bergen, at what is now called James J. Braddock Park.

It was such a clear day; you could not mistake this object. We were so amazed, we thought this would be in the newspapers, we thought a lot of people would have seen it. There was no mention of it anywhere in the media. Altogether there were four eyewitnesses. I could not believe that such a bizarre object could fly over such a densely populated area, without a lot of notice and some kind of commotion happening. At this point I realized that people were extremely unobservant in this area, which is the suburb directly across the river from midtown Manhattan. This object had obviously flown over midtown Manhattan before we saw it because it maintained a straight flight path from east to west. It did not resemble any type of aircraft and it had nothing protruding from it, no sound. This thing made no sense; we were shocked at what we just saw.

Now that I think back on this sighting after many years of research, I believe this object was some kind of a remote-control atmospheric probe. I've never seen anything like it again. The fact that this occurred over such a densely populated area without causing a mass sighting and a mention of it in the media was totally mind-boggling to us. Once again I realized that people in a city area were extremely unobservant of the sky. I'm sure that somewhere in the area someone else saw this object. One day I'm sure I will hear from them. But the four of us will never forget it. I thought to myself, 'people in this area are asleep at the switch'.

This day began my new habit of constantly observing the sky nonstop day and night. The movement of this object could be described as tumbling forward in a circular motion in a clock wise fashion rotating towards 12:00 o'clock, then 1:00 o'clock, then 2:00 o'clock, 3:00 o'clock, and so on. The movement of the object was more bizarre than any sighting I have had since. This was my first sighting. After that day I became a much better observer, I began to learn how to estimate the size and speed of aerial objects and also distance, shape, lights, color and direction.

Black Metallic Cylinder

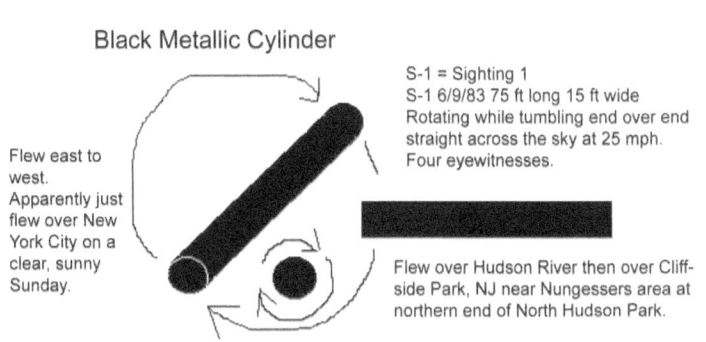

Flew east to west. Apparently just flew over New York City on a clear, sunny Sunday.

S-1 = Sighting 1
S-1 6/9/83 75 ft long 15 ft wide
Rotating while tumbling end over end straight across the sky at 25 mph.
Four eyewitnesses.

Flew over Hudson River then over Cliffside Park, NJ near Nungessers area at northern end of North Hudson Park.

My second sighting

1988

Five years after my initial sighting and now living in Gardiner N.Y., we had moved upstate to a house with some nice acreage. Across the road from us were hundreds of acres of land. There were fields with hundreds of acres of forest behind the fields. It was March 10th, 1988 on a Thursday night a little after 9:00 p.m. My father and I were in the living room in the ranch. It was an extremely dark night with no moon out. My father looked out the bay window in the living room and asked, "What the heck is this?" His first impression was that this was a C-5A cargo jet with lights on it that was about to crash into the woods.

I looked out the window and saw large round colored lights in a huge pattern moving as one, above the treetops, across the road out over the forest, about a quarter mile away. We both went out on to the front lawn looking at this object. I saw round colored lights, some very big and some smaller. These lights were attached to something gigantic. It was at least 1,000 ft. long. There were at least 40 lights, they were all different colors, and the colors were very pure as if created by light with a prism. These lights were much like Chinese globes, they cast no beam forward and they did not illuminate the craft. You couldn't see the body of the craft in the darkness, as though it were black or non reflective. You only saw the colored lights on the craft. It came out on an extremely dark night that hid its body shape. At the far end of the craft was a single red light that was very tall, shaped like a standing cylinder. I said to my dad, "Where is the sound?" We listened for a few seconds and noticed it was completely silent; it was floating by very slowly, perhaps only 25 mi. per hour. It was obviously one giant craft, all of the lights were perfectly situated while it moved, it was confusing to look at first, but we recognized it was one giant craft within seconds.

It slowly floated over the treetops perhaps only 50 ft. above them. This huge thing slowed down to a hover. It hovered in one spot for a few seconds, and then it rotated on its own axis, and turned South East and floated down towards Stewart Air Force Base. This is back when Stewart was a military base. There were no commercial flights back then; it was fully military at the time. The population in the area was much less at the time. It slowly floated away with the large red light at its rear end very visible. That tall red light encompassed the height of the craft. I estimate the height of the craft as about equal to a six or seven story building and the length was roughly 1000 feet.

Afterwards my father and I were talking and estimating size, distance, and speed. We talked about the bizarre lights with perfect pure colors, and a lack of any sound. We could hardly believe what we saw and yet we new exactly what we saw, because we are both very good observers. A month or two after this we turned on the TV and saw W T Z A out of Poughkeepsie, they were discussing the Hudson Valley UFO, in the report by Tony Fama, and he described thousands of area residents seeing this ship.

In this report they showed the video clip of the Hudson Valley UFO taken in Brewster, N.Y. by Bob Pozzuoli back in 1984. My father loudly proclaimed "that's the thing we saw over the forest in March"! We knew that was it when we saw the tall red cylindrical light in the rear of the craft, also that there was no sound captured on the videotape.

After that we bought a book called Night Siege about the Hudson Valley sightings and saw the various drawings, and various light patterns the craft displays. The light patterns the craft displayed that night were the same as when it appears in its boomerang shape. It became apparent to me that this craft displays different patterns of lights such as the circle of white lights with the large red light trailing it, as in the Pozzuoli tape, and also the round colored lights going up and down the side, such as when we saw it in March.

We thought about the fact that it comes out on pitch-black nights and shows you whatever pattern of lights it wants to. This technique hides the true shape of the body of the craft. The craft acts as a computer display screen, and confuses you as to its real shape. That was the big ship. It was an incredible thing to see. It was the biggest machine I've ever seen in the sky, and I was sure that I was seeing a machine, a computer controlled anti-gravity machine.

After this, my second sighting, I became super observant; I was on red alert after that. I became much better at estimating size, altitude, distance, sound, and speed. It was a little too close to our house for comfort, only a quarter mile away. This began the incredible UFO flap of 1988 in this area. After that I was becoming very exacting in recording the date and time of any future sighting. At that point I had not yet discerned the true shape of the craft, but this was to come later.

The day after that sighting 3 unmarked black helicopters flew over the flight path the ship took. The helicopters flew in circles looking down on the area where the huge craft hovered for about 30 seconds, then they flew the opposite flight path the huge craft took, retracing its flight. The helicopters were extremely loud, the ship made no sound at all.

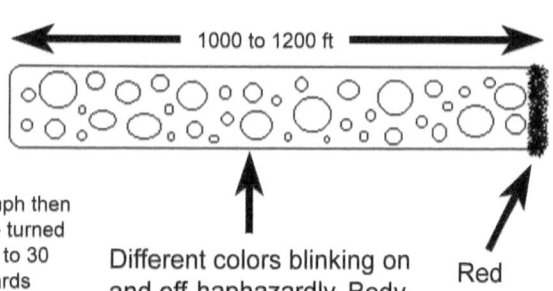

1000 to 1200 ft

S-2 3/10/88 30 mph then slowed to hover - turned on its axis - back to 30 mph -- went towards Stewart Air Force Base - Newburgh, NY - 2 eyewitnesses - many more in area and extended region.

Different colors blinking on and off haphazardly. Body not visible, only lights.

Red

Very Dark Night

My third sighting,

This was my first close encounter, occurring on Burnt Meadow Road in Gardiner. Back then I had a Kawasaki 125 dirt bike I used to ride through the fields and forests in the area. I knew plenty of landowners and got to meet a few more. I rode up to a small hilltop that gave me a good view of the surrounding area. I turned off my motorcycle and I sat on the seat enjoying the beauty of the forest in that area. I had my audio cassette player, a small pocket type player; I put on my headphones and was analyzing some awesome guitar playing while enjoying the beauty of the forest.

I began to see something coming over the treetops only about 300 ft. away. At first it looked like a V formation of birds flying by, but as it came closer I could see it wasn't birds. It was a V formation of grey squares, as it came closer I saw grey lines almost like ribs coming down from each square, as it was coming closer to my view I could now see the outline of a large craft. What I saw was like grey shadow lines. Almost like something that was drawn with a light colored pencil. These light grey lines outlined the shape of the craft perfectly. This craft was flying directly over the Galeville substation, which is a large electrical transformer where power lines meet in the area; this was a close encounter because the craft passed by only about 200 ft. from where I was at its closest approach. The shadowy lines outlined a diamond shaped craft.

The squares were along the top edge on its side and the ribs came down from each square. I was seeing a craft that was 95% invisible and I could only see these light shadowy lines outlining its shape. At this point I had taken my headphones off and I could hear no noise coming from this thing. It passed over a large house in the field next to me, I know the son of the man who owns that house. As it passed over I could see the underside of the craft, and I could see details in this structure and levels going up into the underside of the ship.

When the ship was at its closest to me, as it passed over his house, with me in the field right next to the house, I felt a brief period of fear because the craft was so close, I felt very vulnerable. I actually waved at the craft knowing whoever was in the craft could see me. Just for that moment I felt fear. It was like a brief fear that only lasted a few seconds. I thought I was feeling some radiation from the craft also. This craft made no sound at all, and it floated by very slowly, about only 20 mi. per hour. It was about 200 ft. long and the most incredible thing about it was that it was mostly invisible, I only saw the light shadowy lines outlining the structure but I could see through the structure. It continued on its path, it went from Southeast to Northwest.

I felt relieved because it kept going away from me, when it was about 300 ft. away from the house I couldn't see it at all. It was only visible within that 300 ft. range. I was shocked at what I just saw. I grabbed my cassette player, put it in my pocket, started up the motorcycle, and rode down to my buddy Dan's house about a quarter mile away, and started telling him what I just saw, and I began drawing a picture of it. In the next few weeks I drew many pictures of it, and then I built a model of it. I now have a couple models of it.

By now I was already researching UFOs and reading some key books. I realized I had seen invisibility demonstrated with this craft. I now believe that the only reason I saw any of the craft at all was because it passed through the electromagnetic field of the substation and that field stripped away some of its cloaking ability. This is the kind of thing you have to see to believe, and I saw it close up, 200 ft. away, it was a scary close encounter. I saw a technology that day that was right out of Star Trek just like the Romulan cloaking device. I wouldn't have believed it unless I saw it, but I did see it. It was so close; I could estimate size, distance, and speed very well.

I now know after much research that such invisibility can be achieved utilizing the large quantity of electromagnetic radiation exuding from the craft, which causes photon streams in the surrounding air at the proper wavelength and frequency that are outside the visible spectrum of the human eye.

The shape of this craft was like a donut stretched into a diamond with lines like ribs coming down the sides with a square at the top of each rib. The underside had different levels going up into the craft kind of like decks on a ship, or like different stories of an apartment building. I estimated size as 200 ft. long and 30 ft. tall. This incredible sighting showed me the shape of a ship that I was to see again at nighttime with its different light formations showing. This sighting occurred at dusk, it was still light out. I knew exactly what I was seeing.

Four hours before this sighting three black helicopters flew the exact same flight path as the craft did; they flew in the same direction at the same altitude almost like a prelude to the flight of the ship. These were unmarked black copters, three of them, just like we saw after the second sighting. Part of the reason I went out viewing that evening, was because I knew that those three black helicopters flew by that afternoon, and that meant to me that there could be another ship out that evening. So I got on my motorcycle and went to my favorite viewing spot on that hillside at the right time.

After much research, and being lucky enough to talk to the right people, I learned things about this craft which have strongly indicated to me what it is, I will talk about this later in the book because it will fit the context better, after you've heard about my other sightings.

By now I was thinking that my life was a freak show because of the incredible sightings I had witnessed. These first three sightings were completely different from each other. After this close encounter I began to get concerned that these things were getting too close to me personally. I began constantly looking out the windows and scouring the sky, observing all the time. I wanted nothing to do with these UFOs besides just seeing them go by. My first sighting there were three other eyewitnesses, my second sighting it was my father and I, my third sighting was a close encounter and I was alone. I was getting militant by now because the second sighting was kind of close to my house, and now my third sighting was way too close to me personally for comfort. I was traumatized, for seven years I slept with a loaded firearm within arm's reach of my bed. I wasn't going to put up with any 'greys' coming to my house or property. We had three cats and a dog at the time and I loved them and was very protective of them also, I was a real magnet for the impossible in the late '80s and early '90s. It was an incredible show in the skies in my neighborhood.

S-3

May 15, 1988

8:30pm est

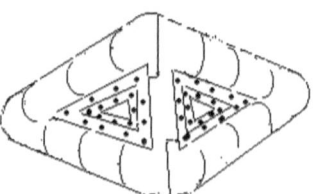

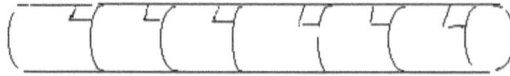

My fourth sighting

June of 1988 on a Thursday evening

I don't remember the exact time or the exact date. I wasn't ready for it. It could've been the 9th or the 16th, but this time I didn't memorize it and, I lost the notes, which I wrote after every sighting. My friend Dan and I were in our back yard, coming up from the river. We had the Shawangunk Kill in our backyard about 200 ft. behind the house. As we walked up to the house, I looked west towards the mountain, which was the face of Millbrook Mountain. I saw an orange ball of light. It was less than a quarter mile away, flying in front of the mountain just above the treetops making no sound. It looked just like a round ball of light of a fiery orange color. It was flying north to south; it was only 30 ft. above the treetops. It was spherical in shape. I would estimate it was about 30 ft. wide, my friend Dan and I were the only eyewitnesses. It moved about 25 mph and fit the description of some craft seen in Russia.

S-4
6/16(?)/88
Orange fireball sphere 25 ft wide moving
25 mph.
2 eyewitnesses just above treeops north
to south.

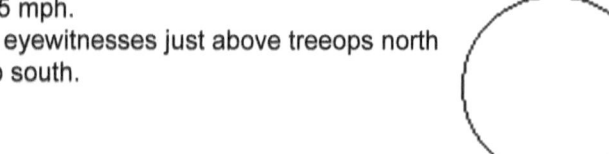

My fifth sighting

Thursday July 7, 1988 at around four in the afternoon est

It was a generally clear sunny day with some clouds in the sky. My cousin Al and I were hiking on the Shawangunk Mountains on what they call over cliff and under cliff pass, which circles the top of the mountain above the hairpin turn. I was looking northeast towards Rosendale, at the northern end of the mountains between a break in the trees, when I saw an object come out of the clouds. It was oval in shape, close to being a saucer in shape. This object appeared to be a brilliant white light as though it were surrounded by ionized air or plasma shield. The object came out from behind a cloud when I caught sight of it.

I saw it perform maneuvers that were impossible for any aircraft. It went back and forth and up and down in bowl shaped maneuvers. It seemed to bounce up and down a little in shaky movements, floated around a few seconds more, and then went back behind the clouds. It was only visible for about 20 seconds. I called to my cousin and he saw it for about the last five seconds. I stood there and watched but it did not reemerge from the clouds, it was almost as if it was putting on a quick performance for me.

It looked just like photos I've seen in documentaries of a classic oval or saucer surrounded by a white ionized air. This again was a unique sighting different from the previous ones, a different craft once again. I haven't seen one like it since. Two eyewitnesses saw this one. I estimate the size of the oval at about a hundred feet across. It was about three hundred feet above the mountain, which puts it at about 1900 ft. altitude. By now I had seen five sightings and apparently five different kinds of craft. I would later find out that the craft in the second and third sightings appeared to be of the same design, one over a thousand feet long, and one about two hundred feet long.

Side bar information: I include this for the sake of chronology of events, so you can get a feel for what was happening at the time, which was a constant barrage of events in 1988. On August 11th my father saw the big ship again in the area. He was driving at the time. He told me he saw it in approximately the same area we saw it on March 10th, just above the treetops.

Side bar info: Thursday night August 18th 1988. By then I knew that most sightings occurred on Thursday night between 9 and 10:00 p.m. I was on the roof of my house with binoculars viewing the area and enjoying the stars and fresh air. I was looking Southeast towards Stewart Air Base. It was far away, but I saw something rise up vertically which was circular with lights around the rim. There were a bunch of smaller lights around it. It rose up a few hundred feet and then went back down, and then the smaller lights flew around the area. It was a flurry of activity. This was over in another minute.

I don't count this in with the sightings that I'm sure of because it was too far away. It happened pretty quickly. It was at least 10 mi. away. It was a very clear night so I could see it anyway. I don't know of anything conventional that is round with colored lights around the edges, which rises vertically, and goes straight back down. This happened right in the area of a military air force base. I leave it as a side bar because I'm not absolutely sure it was an unconventional craft, though I believe it was. It was too far away from me to be positive, so I don't assign a number to this event. I only number them if they're close enough for me to be sure of shape and silence.

S-5
7/7/88
Plasma light - oval - 100 ft long
2 eyewitnesses

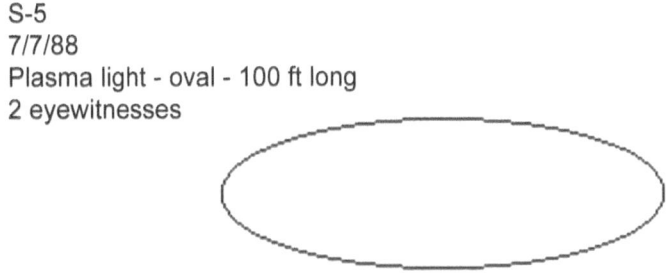

My sixth sighting

Sunday August 21, 1988 about 10:45pm

This occasion was my second close encounter, my first being on May 15th. I was returning from my cousin's house in Connecticut. I was pulling up to my driveway in my car when I looked in the field across from our house. I saw something floating over the field that had two large round white lights in front. It was about 400 ft. away from my car. It was only about 20 ft. off the ground. I immediately rolled down my window and listened, and heard no sound. I shut off the lights on my car and sat there in the road for a minute looking at it. There was no traffic at all on my road back then.

It was moving at about 10 mi. per hour like the speed of a man running, perhaps slower like jogging. It was slowly making a turn, right at the point when I shut my lights off; the two white lights on the craft were pointed directly at me. When the lights were pointed straight at me they were extremely bright. I looked at them but I had to squint. My mind went blank for a few seconds and I felt a slight fear of vulnerability for the two seconds the lights were on me. The craft continued on its slow turn a second later when the lights were not pointed at me. I could see that the lights cast no light forward. They were like Chinese globes. They were bright but they did not illuminate, yet, when they were pointed straight at me for those two seconds it was like being in a beam of light. They were extremely bright when I was straight in their aim; it was like being in a tunnel of light extremely bright making me squint from 400 ft. away. But the second they weren't straight at me, they seemed so much dimmer with no light coming forward. I think this is subatomic particle beam technology at work, another thing someone stole from Tesla.

They were two round white lights about 6 ft. tall. At this point I was out of the beam, and the craft was continuing its turn. I turned on my lights and sped down my driveway that was about 200 ft. long. I ran into the house, my father and my friend Dan were in the house, I yelled to them, 'The small ship is in the field across the street'. Then the three of us ran up the lawn to the road, by now the craft had completed its turn and was flying away from us barely above the treetops. Now what we saw was lights on the underside of the craft in the shape of an arrow, they were small red lights in the front of the craft, and there was a tall red and blinking light in the back, looking just like the one on the big ship. It silently floated down towards Stewart AFB, or in the direction of it. This craft was barely 15 ft. off the ground when the two white lights pointed at me, in complete silence. When I saw the dim red arrow of lights in the front on the bottom, and the red light in the back, as it flew away from us, I realized that this was the same small diamond shaped craft that I had a close encounter with on May 15th because it was the same size and shape.

When this happened, I had just returned from my cousin's house in Connecticut where we were discussing sensitive UFO information. I had just given him an article in an old copy of Argosy magazine by Renato Vesco, who was an aerospace expert who worked on ramjet development. He worked with the Nazis before World War Two. He talked about the **foo fighters** being a secret German weapon called the **feuerball**, or fireball; he said it was a remote-control weapon which brought down our bomber planes over Germany by short circuiting their ignition systems by exposing them to dangerous electrostatic fields. He said the first flying saucers were made by scientists in Nazi Germany. Al wrote articles on Paranet about this in 1988. Vesco wrote a book called **Intercept But Do Not Fire Upon** that was also quoted in an official Air Force release to pilots. Vesco is one of my many sources of information in my research during and after my sightings, I shouldn't say after because they are still going on, only now they are fewer, and further in between. I will discuss my research in a later chapter, so I can continue describing my sightings first.

S-6
8/21/88
6 ft tall - white lights in front - moving 10 mph

Side to underneath view all lights red seen from
behind moving away - small ship 200 ft. long

My seventh sighting

September 8, 1988

This one was the most obvious to all. It was again a Thursday night. My cousin Al and I wanted to view the sky to look for UFOs. We had a weird feeling like we knew we would see it this night. We went up to the hairpin turn on Route 44-55. The ship was already out waiting for us. It was the big ship, the same one as in my second sighting, about a thousand feet long. It was at an altitude of about 1,500 ft., about equal to the height of the parking lot above the hairpin turn. It was 9:10pm.

It was floating around extremely slowly, no sound as always; the light pattern was exactly the same as on the videotape taken by Bob Pozuolli in Brewster in 1984. There was a circle of white lights on the front part of the crafts underside. These white lights seemed to slowly rotate. There was that old familiar tall cylindrical red light blinking at a steady interval on and off. My cousin was freaking out because he had never seen a UFO before. I was calmly staring at it and analyzing its every move, by now I was used to seeing UFOs, after what I had already seen. I reproved him for doubting my sightings in the past, because he had made fun of me in the past over my sightings. Now he was seeing the big one in his face. We had our binoculars with us. We watched closely for 20 minutes, we drove up to the parking lot that was 200 ft. higher in altitude, a short distance away, and watched it for another 15 minutes, through astronomical binoculars.

The body was just a black mass blocking out stars. At a certain point the red light in the back appeared to be blink out, then shoot up to the front and spin around the white lights three times and then shoot back to its original position. From the ground underneath it , this would have appeared to be a group of small lights, or small craft, with the red one moving at incredible speed, around the other's, but we were at the same altitude as the craft, and could easily see that all of the lights were on the craft. The craft acted as a huge computer display screen, with digital precision.

Then the light pattern changed. All the lights went out, and then a new pattern of light came on. The new pattern was an arrow of red lights in the front, with the large red light in the back still blinking at a steady interval. All these light changes occurred on the underside of the craft as it floated around at 1,500 ft. I knew that this pattern I was now seeing was exactly the same as the pattern on the small ship on August 21st. This made me think that the big ship was a big diamond, and the small ship was a small diamond, same design. The same light patterns, same digital precision, the same silence, a large antigravity ship, and a smaller one appear to have been made.

Then all the lights went out again. Then the lights came on with a new pattern, the new pattern was typical of the boomerang shape. Now the lights were on the sides of the craft and the bottom, in front. With the big red light in back still blinking that was like a tall cylinder, glowing red, blinking on and off about two seconds on and two seconds off. This new light pattern was the same I had seen on my first sighting of the big ship March 10, 1988. Big round colored lights with the smaller ones around them.
Pure prismatic colors, but no projected illumination, like Chinese globe lights.

This is the same light pattern a lot of people saw in the early '80s. The same pattern is on the cover of Night Siege, the book about the Hudson Valley sightings by Dr. J. Allen Hynek and Phil Ambrogno.

My cousin and I drove back to my house because the craft seemed to be in that area. We got to my road; and watched it for a while, and found two of my neighbors watching it on their lawns. We watched it for about 35 minutes in total that night. This craft was visible to an enormous area that night. Anyone from Newburgh to New Paltz could have seen it that night. A lot of people did, the state troopers telephone switchboards were jammed with callers. You couldn't even get through to them on the phone. My next-door neighbor was a retired state trooper. We asked him to get the scoop.

The official explanation to the public was ridiculous. They said it was a hoax perpetuated by a group of ultra- lights. Some of the round lights on this craft were as tall as a five-story building. The big red light in the back of the craft was as tall as a seven-story building. It was one gigantic craft with lights on it. It made no sound whatsoever. It flew as slow as 25 mi. per hour. It floated in the air, and it rocked up and down. It was getting sideways in the sky at a certain point, with the rear end dipping down low. It was obviously an antigravity machine. At a certain point a small airplane approached it, it got close to the craft, and all of a sudden did a 180 degree turn, very quickly in a super tight circle. It was as if a big invisible hand had spun that plane around.

What I saw that night was a ship that defies gravity and displays lights exactly like a computer display screen with digital precision. It was at least a thousand feet long and was absolutely silent. While this big ship was floating around the area for all to see, people later reported that the small ship was floating around Wallkill and visible from the Thruway market area at the same time. A huge ship visible from a huge area for over a half hour and no mention of it in the media, of course I knew long before this there was a huge cover-up going on. I knew that on my first sighting. The cover-up is so huge it's worldwide. All UFO researchers know this if they know anything at all. Later I found that local newspapers had mentioned this event, but with little accuracy. They only mentioned the small craft in Wallkill. I thought the sighting was big enough to be worldwide news.

This sighting was the biggest and most obvious of all the ones I've had, the only camera I owned at the time was a Polaroid 600 that is only good during the day for a short distance. I owned no cameras besides this, and I knew nothing about handling cameras. The one neighbor who had a 35 mm lived in the valley by the river up the block. He had no view because he was low in the trees where his house was. I couldn't afford to go buy cameras. I wasn't thinking about photographing the ship because they were constantly out.

Anyone who made a wise guy comment to me got the same answer. I told them to hang out with me by my house every Thursday night and you will see one. The few people who took me up on it saw one. The eyewitnesses I know for this sighting included my cousin and father and 4 neighbors and many people they knew in surrounding towns and many other people I heard about. My neighbors and I have spoken to many other eyewitnesses in the area. I heard other people talk about the UFOs in my area before I saw them. This was a big sighting. My cousin described it first on his computer in Connecticut and this led to him and I being identified as verified eyewitnesses to this sighting. Phil Ambrogno noticed we were the first to accurately describe it. He gave our names to Good Morning America and we wound up being interviewed and appearing on the show briefly.

1988 was such a busy UFO flap, that you didn't need a camera, we didn't think much about cameras. I only thought about constantly watching the sky so I could catch the next ship and call my neighbors to get more eyewitnesses, which happened many more times.

After this sighting, my cousin Al wrote a bunch of articles on Paranet when he lived in Connecticut. They had titles like UFOs Alien or Man-made?, and Cyborg UFO. After his articles, he became a much talked about mystery on Paranet. He got his direction from me. I led him to the right books and articles to help figure out this phenomenon, He dug up some good information too. I will talk about my findings later.

My cousin had one big sighting and it blew his mind and made him write articles. He did see the white oval for about five seconds also. I had 14 sightings so far, I'm writing this particular segment on December 22, 2006. I'm now a unique UFO researcher and I have some information no one has heard yet, I will discuss these things in a later chapter. I will continue describing my sightings, and discuss the findings of my research later.

Thursday night October 13th 1988, 10:00 pm est

UFO hoax, airplanes flying in formation over the Hudson River. The planes were dropping flares, along with some fluorescent material that appeared to be liquid, it was a strange sight, it was obviously none of the UFOs. It was about 15 mi. away as the crow flies. But I saw it well because it was a clear night. I looked at it through binoculars, the lights moved in relation to each other, definitely no ship. I think the military was playing with our minds in order to confuse people to help cover-up the real ships.

Nov3, 1988

My cousin Al and I were interviewed by the science editor for Good Morning America, Mike Gilliam, I believe was his name, though I'm not sure of the spelling of his last name. We met with their camera crew up by a hairpin turn, where we had seen the big ship on September 8th. I got to meet some other eyewitnesses of the big ship. I met Phil Ambrogno, the co-author for the book Night Siege, I also met Dennis Sant who had a close encounter, and once appeared on Unsolved Mysteries, and they were very interesting gentleman. We described what we saw that night to the film crew; we were interviewed for about two minutes each. When the television show came out, we had been edited down to about five seconds each.

Thursday December 8, 1988, 7:22 pm est

My father saw what looked like the experimental military aircraft nicknamed the **pumpkinseed** go by at high altitude he estimated it at about 30,000 ft up. It was difficult to estimate the incredible speed it was traveling at. He said it went by so fast you could barely see it. But he was able to make out the shape. This aircraft had been briefly referred to on a cable television show at the time about new high-speed military aircraft.

All sidebars are included to show you the chronology of events, as you can see this was a busy year.

Thursday. April 27, 1989 9:50 pm est

My neighbor Dawn across the road saw something in the field next to her house and caught it on a camcorder for a minute. I have the film, it was taken at nighttime, and you can see small white lights going on and off in different places in a set space, but you don't see the craft. I believe she saw the small ship with its cloaking device on with the small lights going on and off around its body. It reminded me of a film I saw once which was produced by M.U.F.O.N. in Long Island which showed white lights blinking on and off on some type of a craft which was silent, in which you couldn't see the body, because it was dark or non reflective, appearing on a very dark night.

That same night hours earlier, four planes flew in formation towards the South, near Stewart A.F.B. I only saw planes flying in formation during the UFO flap of 1988. I also came to see later, that the small ship could display lights, white ones, while being cloaked. I saw that on my tenth sighting which we will get to soon.

Saturday May 20, 1989 10:55 pm est

I went out on my back porch to feed the cats. I saw something 1/4 mile away that looked like it could have been a small ship. I called my father to come out and look. It was a group of small lights traveling together, there was a humming sound, which reminded me of the sound of ultra-lights, they were quiet for ultra-lights, but that's what I think it was. The lights moved in relation to each other. I think they were set up with lights just to confuse people.

This was a high-tech feat for a group of ultra-lights. They did a pretty good job of flying in formation, and I don't know how they got those lights on them, but someone funded a hoax that night I believe. Every ship I've seen had no sound, and most came out on a Thursday or Sunday night. Except my first sighting that was a Sunday during the day, and the one with the white oval which was a Thursday during the day. My first C E -1 was at dusk, but still light out. This next one also broke the Thursday- Sunday rule.

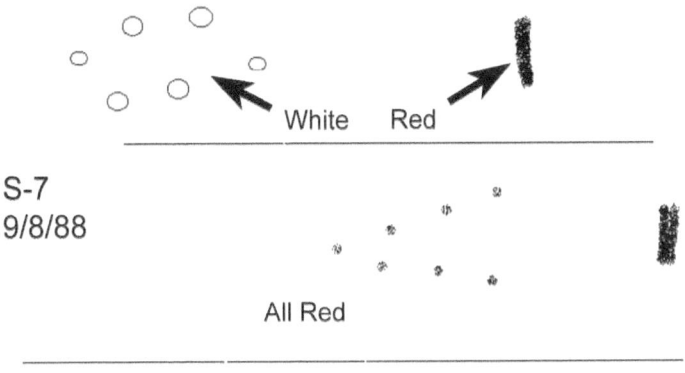

S-7
9/8/88

All Red

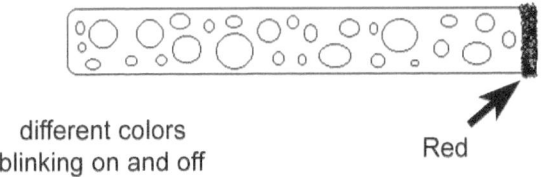

different colors
blinking on and off

Red

Big Ship - 1000 to 1200 ft long

My eighth sighting

Tuesday August 8th 1989 11:14 pm est

I was alone for this one. I went out onto the front lawn to view the sky, and get some fresh air. I was looking up scouring the sky, when I saw the big ship zip by at tremendous speed. Its lights were completely out, then a light came on illuminating the underside, it was bathed in a yellow-orange light. I could see it was the diamond. It shot across the sky in two seconds. I saw the light go out on the other side of the sky. It was traveling northwest to southeast, which was the traveling route most of the time for the diamond. I would estimate its altitude at close to 10,000 ft. Like similar craft observed in the past, the speed was beyond anything observed that was conventional, either commercial or military. It seemed as though it blinked on just to give me a show.

The timing was incredible that I happened to be on my lawn looking up when this happened, it only lasted two seconds, but the light blinked on, and showed me the shape right where I was looking. This was a classic example of an aircraft violating FAA rules. No lights, then one light, then no lights again. I barely caught it, but I saw the shape. Once again no sound, though it was too high up to hear anyway even if at had a sound. That was the fastest I've ever seen anything move in the sky aside from "shooting stars" in the atmosphere. It was very high up, but covered the whole sky in two seconds, and we could see a large piece of sky there.

Sidebar info:

October 14, 1989

The stealth fighter flew in our area around Gardiner from the direction of Stewart AFB. It had a very low-pitched rumble; it was the loudest jet I've heard next to the Concorde, but with a very deep low sound. It had a very unique sound, unlike any other jet I've ever heard. It had a single jet nozzle in the back. It was circular and large, and an orange flame could be seen. My father knows much more about aircraft than myself. He said it was the stealth fighter. Small fast and loud, with a low-pitched sound you feel in your chest. Very much unlike the stealth bomber, which is large and quiet.

This jet flew in our area a few more times; I recognized it immediately because of the low sound, and the single-wide jet exhaust opening in the rear. I didn't write down the dates, It wasn't very interesting after seeing ships. After a period of time I didn't see that jet anymore. I'm sure it was invisible to radar, but you sure can hear it, speaking of radar, most of my sightings were invisible to radar anyway because they were just above treetop level, flying below radar.

S-8
8/8/89

Big Ship
10,000 ft up

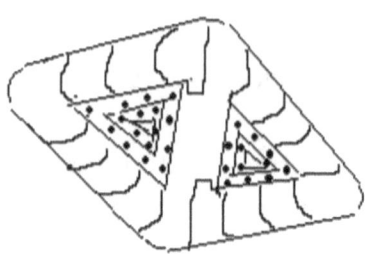

My ninth sighting

Thursday November 9, 1989 5:55 pm est

I was in my bedroom when I heard the familiar sound of a C-130 turboprop military transport plane. It made me look out the window. I saw the C-130 heading down towards Stewart. Something like 400 ft. behind it was what I believe was the small diamond. There were three white lights in the shape of a triangle on the craft. It had the familiar floating and rocking quality that these things have, and of course I heard no noise other than that C-130.

Apparently that plane was escorting the small ship in. A half an hour later, we saw of C5-A cargo jet, which is the biggest we have, they had a bunch of them at Stewart at the time. It was flying circles around the area and it had three white lights in the shape of a triangle on its underside. It was clearly simulating the lights on the small ship. Someone was making the effort to have a cover story when the ships come out. The three white lights on the C -5A projected light off of them; they looked like some kind of a normal light. But the three white lights on the small ship had that abnormal quality about them. They looked weird. They projected no light. The lights on these ships looked nothing like any other lights you'll ever see. I saw the ship with the C-130, later my father saw the C-5A with me. One of my neighbors around the block, Melinda, saw the ship behind the C-130 also. She reported exactly what I saw. She knew that wasn't any normal aircraft behind the C-130.

S-9

11/09/89

Small Ship 3 white lights on underside 15 ft wide 200 ft long ship 300 ft
behind C-130 Turboprop military
transport plane heading towards
Stewart AFB

My 10th sighting

Monday January 22, 1990 5:30 am est

My father was sleeping on the couch in the living room. We both awoke to the sound of a C-130 turboprop, my father yelled to look out the window. My bedroom window and the living room bay window both faced south, looking out over the field where we first saw the big ship. I yelled back to him "I see it". I saw the C-130, and behind it was the small ship. I saw the big round white lights in the front, just like on my second close encounter. Now it was light out in the morning, it was barely the light of day-break. You could not see the body of the craft. The fuselage was invisible. You just saw the round white lights in front. The white lights followed the C-130 and headed north.

I ran into the living room. Dad and I described the same thing. Dad got to see the effects of the cloaking technology that morning. I described it to him regarding my first close encounter with the small-diamond. But this was the only time he had seen it. When we caught sight of the plane with the white lights behind it, it was only a quarter mile away. It came in a direction right over our house. Our windows faced south.

In that direction was not only Stewart A.F.B., also the Galeville Special Operations Center, which was a military base that apparently had underground facilities. Parachutists were trained above ground, and other types of military training. The area was surrounded by trees and very difficult to see into. It was right where Gardiner borders Pinebush on Hoagerburg Rd. I knew a neighbor who once told me that he saw the small ship operating in that area at treetop level. Pine Bush is more famous for UFO events and Pinebush borders Gardiner to the south. We were looking out over Pinebush and the Special Operations Center, and Stewart AFB in those years from that house. We also saw a large piece of sky where we were.

The Special Operations Center is now a park called the Shawangunk Wetlands. I used to drive up to the entrance of the Special Operations area and read the sign. It said that entrance into this area without prior authorization from Army, FBI, or CIA was strictly forbidden and violators were subject to immediate arrest. I don't remember what the sign said it in its entirety, but that was the important part. Over the years I've brought some friends to see that sign. There was a camera pointing at the entrance, it was inside a box of some kind. Now that it's a park, of course the sign and camera are gone. Things sure changed when Stewart went commercial and was no longer just a military base.

Sidebar info:

March 13, 1990 9:30 pm est

Television show Beyond 2000 showed a fusion reactor made in Japan shaped like a doughnut with ribs going down the side, it was the tocomac style reactor, and there was talk of solving all energy needs with one source. This will come up later in my research.

S-10
1/22/90
Small Ship no body visible - white lights
15 ft wide - behind C-130 turboprop
coming from direction of Stewart AFB
headed north-northwest

My 11th sighting

Thursday April 5, 1990 9:50 pm to 10:00 pm est

The small diamond 300 ft. up, going north to south. It was east of us by about a mile moving slowly, like 40 mph. This time I saw the tall red light at both ends of it. In between there were at least 50 lights of different sizes and colors blinking on and off, up and down the side of the craft, the lights were round. At least four of my neighbors saw it with me because I called them on the telephone after seeing it and running inside. It was visible for a good five to ten minutes.

Sidebar info: Thursday May 31, 1990 9:55 p.m. UFO hoax, a group of lights out to the East flew in formation and then broke up. Some of the lights blinked out completely, again violating FAA rules. I know at least six neighbors who saw this one. We knew it was aircraft jerking around. Once again someone was trying to confuse area folks. It wasn't working.

Sidebar info:

Sunday January 21, 1991 9:27am

There was a snow squall going on, I was lying in bed awake, trying to go back to sleep for awhile, when all the sudden there was a huge explosion. There was a big shock wave that rippled through the ground and shook the house. It was the biggest explosion I've ever heard. A wave of brilliant white light passed right through me, I saw it with my eyes closed, everything lit up white, it was like a wave of high frequency radiation. It reminded me of a nuclear blast of some kind, the sound of this blast was heard as far away as New York City, probably further from the sound of it.

We spoke to other neighbors and friends in the area, who said that the sky lit up white when that happened. More friends in the area said that they saw it through their closed eyelids, a brilliant flash of white light penetrating right through you. I saw a quick blurb in a newspaper a few days later, It said that an unusually strong thunderclap had occurred during the snowstorm, and the sound was heard in New York City, coming from the Hudson Valley.

What I have to say about this is, it was not thunder, and it was one huge explosion. I've never felt a thunderclap shake the ground like that. I've never had a white brilliant light penetrate my body and my eyelids like that, and there was no other thunder or lightning in that storm, it was unique. I speculated that the big ship may have blown up. I'll get into why I think that later in my research chapters. After that I never saw the big ship again. I only saw the small one and a medium one.

The newspaper referred to it as a sky quake. It could have been a beam weapon type of thing, or a piece of anti matter from space exploding, or the big ship exploding, but I'm sure it wasn't just a thunderclap or a "sky quake" what ever the heck that is. The sound and the shock wave were amazing, but it was that white light, like plasma, passing through my closed eyes, which was abnormal, I've never experienced anything like that before or after.

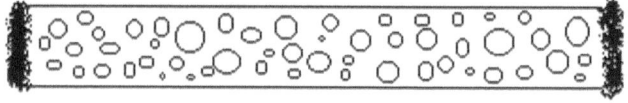

S-11
4/5/90
Small Ship - multicolored lights blinking on and off
- red ones on both ends

My 12th sighting

June 6, 1991

It was a Thursday a little after 9:00 p.m. I would call this one a medium-sized diamond. This time my neighbor called me up to see it. He is a retired state trooper, it was his son's birthday, it was about 2,000 ft. up this time, and it was traveling west to east. It had just come over the Shawangunk Mountains and went right over our houses, ours and our neighbor who called. This one was about 500 ft. long and was traveling at a couple hundred miles an hour. It had the tall red light at either side of it, front and back. It had at least 75 lights on it; they were round, different sizes and colors, blinking on and off like crazy in different patterns up and down the sides.

I was doing construction at that time, I was a masonry laborer, I went to work the next day and my coworker who lived in Poughkeepsie told me that a space ship floated over Poughkeepsie last night. He told me it floated right over the IBM building, and over the hotel he was staying in. He described it to me in detail; it was the same craft, at roughly the same time. When we saw it, it was heading east. I knew it would be over Poughkeepsie in a few minutes. Tony had never seen anything like it. He was completely amazed. It was silent, floating and rocking with blinking colored lights all over it. I was used to it by now. The craft used the same light patterns as it did on April 5, 1990 on this sighting. But this was the first time I was sure that it wasn't the big one or the little one, it was about a 500 ft. machine, same design though, the same lights and light patterns, shape, and silence. It also had that typical floating, rocking quality.

Sidebar info:

Monday December 14, 1992 2:00 am est

The hum; For the sake of chronology I include this, it may be related to UFOs, or it may be related to underground tunneling, or it could be an experimental device using sound waves on the public while they are sleeping.

I woke up at 2:00 a.m. I was awakened by a sound that was a deep low frequency hum, it was very quiet but I could hear it. I was lying still in bed; it was perfectly timed intervals of low frequency hums. At the beginning and end of each interval the frequency was a tiny bit higher. It was a super quiet night, so I could hear it well. I realized for the first time that I had heard this sound before, late at night, when everything was quiet. It sounded like it was coming from the south near the base of the mountains, in the Pine Bush area, or perhaps in the area of Walker Valley or the town of

Shawangunk, which borders Gardiner to the southwest.

I felt that my subconscious mind had woken me up so I could consciously hear this sound; I know I had heard it before on other quiet nights. But this night I woke up in amazement knowing I've heard it before. I discussed it with my father and he told me he had heard it many times. It happens in the late morning hours when everyone's sleeping. I heard it again a few months later at about the same hour when I left my friend's house down the block to come home. Not long after that my father got a book about underground tunneling machines. He thought that is what might be causing it.

I heard Dr. Nick Begich refer to this as the Kokomo hum. It has been heard in different places. I think there's a good chance it comes from underground tunneling machines. I think the mountains here are a great place to tunnel and mine because of the immense mineral wealth in these Shawangunk Mountains. There are also a lot of quartz crystals in these mountains. This also pertains to my research. Crystals can be used to synchronize computers, they are also batteries, and they could have incredible uses for high-tech machinery. I understand there are also light metals in these mountains that can be used to make alloys... very interesting. What a great place to do some secret mining.

In 1994 we moved from that house, we lived down by the Tuthilltown Gristmill for four years, we had no view of the sky down there, and had no sightings there. But our friends on Sandhill Rd. had a great view of the area, and they saw the small ship a couple of times after we had moved. In 1999 I moved over the mountain to Kerhonkson on route 4455 where I live in an area high up on the mountain above 900 ft. in altitude, on the western side of the mountain, before I was on the eastern side. This brings me to my next sighting.

S-12
6/6/91
Medium Ship - 500 ft long

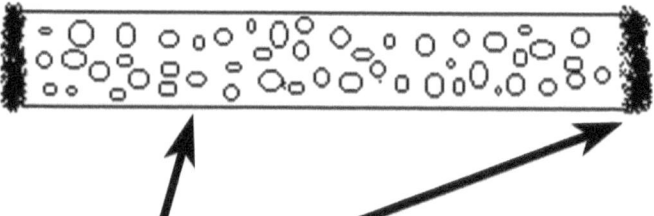

Blinking multicolored - Red on both ends - Same as S-11 but larger

My 13th sighting

Sunday February 24, 2002, 4:37 am est

I was staying up late in the winter listening to Coast to Coast AM and reading books and playing some guitar. I stepped out onto my back porch with a feeling that I should view the sky. I looked towards the mountain top, which is less than half a mile from me, more like a quarter mile and change. It was absolutely silent that morning, no wind and no rustling of trees. I saw the small ship, which is what I usually call the small diamond. It was barely a quarter mile away, it was absolutely silent and had the colored lights up and down the side, I recognized it right away because no aircraft that big can fly that slowly.

This thing floats around real slow, and it's as long as a commercial jetliner. Even an ultra-light would've made a loud racket at that time of morning in such silence, but it was completely silent, I knew it was the small ship, and it was headed south. It was flying in the area of Stony Kill Falls that has a gravel pit near it. The craft was just above the treetops, between myself and the mountain. The altitude where I am located is little over 900 ft.. It was the first time in quite a while that I had seen it, I wondered if it was following me over the mountain.

S-13
2/24/02
Small Ship moving
30 mph - 200 ft
long - multicolored
round lights only

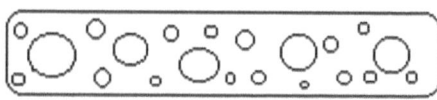

My 14th sighting

Wednesday October 20, 2004, 8:25 pm est

I was on my way to Rosendale to record guitar music at a studio; it was my first recording session at a pro studio. I was recording some killer solo acoustic guitar material. I've been playing over 20 years.

Anyway, I was headed north on Lucas' turnpike, when I saw what had to be the small ship between myself and the mountain, it was floating above Clove valley, which is down below the Mohonk Hotel. This was in the town of Alligerville. It had the old familiar pattern of three white lights in a triangular pattern, once again very slow at about 20 mi. per hour, absolutely silent, you couldn't see the body, and the lights were bizarre like white Chinese globes. They seemed to have lights inside them, smaller lights, almost like some kind of third dimensional Christmas ball. This was only 20 ft. above the treetops. When I intersected where it was and passed it, it was only about an eighth of a mile away tops, maybe as close as 500 ft.

At this time, before 830 pm, It was the opening pitch of the final World Series game between the Yankees and Boston. Hardly anyone was on the road; everyone was in watching the deciding game. I was thinking on the way to the studio that my old antagonist was here bothering me again, trying to make me nervous before my recording, but that didn't stop me from making a good recording.

Sidebar info:

Thursday February 23, 2006 4:15 am est

I was up late during my Coast to Coast AM hours. I went out on the back porch, I just had that feeling to go out and view for a minute, and of course to catch the fresh air and the beauty of the surrounding forest at this quiet hour. I saw what looked like the small ship in the same spot as it was Sunday February 24, 2002 at 4:37 a.m. It was at treetop level and silent, heading in the same direction as last time, I only saw it for a few seconds then it was out of sight. There was little bit of wind that morning, rustling the trees a bit unlike the previous sighting from my back porch, where it was super silent. I won't assign this one a number, I didn't see it long enough, I'm not positive if that was the ship or not, but jets aren't silent, this thing was. The fact that it was a Thursday morning made me think that it fits the schedule.

Saturday morning, December 30, 2006. 3:35 am to 3:50 am est

Up late during Coast to Coast AM hours, I went out on my back porch and looked around just in time to see a light flying over the treetops. It was maybe a quarter mile away in front of the mountain, about 200 ft. below the top of the mountain. It looked like something relatively small with a white light on it, headed east.

As I watched it, it curved upward and then back ward vertically and around in a spiral forming a complete circle and then another circle inside of it.

I've never seen anything move that way in the sky. This movement seems impossible for anything but an antigravity craft. No helicopter could do that. This wasn't a helicopter, it made no sound, and if it was a helicopter it would have been extremely loud especially at that time of morning. The sound would have echoed off the mountain back at me. I went back in and grabbed my binoculars. I came back out and saw the object move to the right a little bit, then sit there and hover. Through the binoculars I could see that there was a few white lights, it looked like three small ones. Also it looked like there was some white light projecting upward from the top of it. It sat there hovering for 10 minutes. I was getting cold, and I went back in to warm-up a little. I came out a few minutes later and it was gone. Maybe I should number this one as a sighting because conventional aircraft can't move the way that thing did. I couldn't see any shape, just a small tight cluster of white lights with no sound. It was extremely bizarre movement.

For now I will leave it as a side bar because I only give a number to a sighting if I get a good look at the shape and I'm absolutely positive that it is unconventional craft. This is a tough decision because I'm sure it was unconventional because of the weird movement and no sound, yet I could not discern a shape, so it's hard to add this one to my other sightings that were all larger and of a clearly discernible shape. This was a unique one and a tough decision on categorizing with my other sightings.

Somehow I have incredible timing with seeing these things, and I go out on the back porch at the right time. I came back in thinking "why don't they leave me the hell alone". After 23 years of observing the skies, I guess I'm a compulsive observer. I can't stop. It's part of my natural habits. I don't really understand how my timing can be so lucky to catch these things. But one of the reasons I've had so many sightings is because I regularly observe, and I learned which nights they were out and continued to observe on those nights. The times were mostly Thursday night between 9 and 10:00 p.m. and Sunday nights from 9 to 11:00 p.m. Remember, people who don't observe don't see anything. Ten out of ten people who doubt you, and say they've never seen anything, do not observe.

This is the end of my sightings and sidebars for now, but I'm always observing and I do live in rural district one of my town. I am bordering the state park with thousands of acres of mountains and forest, so I'll keep looking.

S-14
10/20/04
Small Ship - Lights 15 ft wide - moving 20 mph just above treetops - body not visible - 200 ft long

Some personal reflections

Anyone who reads this book should be able to tell that nobody can make this stuff up. I would have to be the greatest science fiction writer ever to come up with these details, but I'm not, and I'm not interested in fiction, because fact it is much stranger, and much more interesting. I am just reporting what I saw honestly. Everyone who knows me will tell you, I have been telling my friends about my sightings since my first one, in detail, and none of the details have ever changed. I have more eyewitnesses for most of my sightings in addition to myself. I know other people in my area who are highly respected professionals who have also seen the ship many times. I called up my neighbors on the phone to get them out many times, and they saw the ships too.

When the book Night Siege came out, it said that 5000 people had seen the ships in the Hudson Valley. Actually by then there were at least 10,000 people who had seen it, maybe more, in Connecticut and throughout the Hudson Valley and other places in the northeast too. By now the number of people must be gigantic. When I first moved to Gardiner, people talked about it before I ever saw it. One time I was in the deli at the mountain store, I overheard two people talking about UFOs. One guy asked the other "have you ever seen any around here" I turned around and saw the guy's face, a look of tremendous fear came over his face, and then he waited for us to leave the deli before he answered the guy. We were going towards the door anyway, and there was no one left in the deli except those two. I knew from his face that he had been traumatized. My father and I have received many stories from locals that are pretty scary. That guy was too scared to talk about it in front of more people. I have a lot of stories I could tell from other people, but I don't have their permission to write about it, so I won't.

People have asked me if I've been abducted. My answer is no. One friend said, "How can you be sure"? My answer is that I have an amazing memory, and I know I haven't been abducted. My connection between my conscious and subconscious mind is stronger than most people. I don't believe my memory can be erased. I don't believe I can be hypnotized either. I have the kind of mechanical mind that can figure out what something is if you show it to me. They wouldn't want to abduct me because I would remember every detail. Everything you show me I will remember and will identify in relation to Earth technology. I'm not the kind of guy they would want to abduct because I would spoil their secrets. After years of research I can look back on my sightings and tell you some amazing things about them, in relation to our technology, or I should say in relation to hidden technology.

Some strong scientific opinions

Like I said, I think the first sighting was a remote-control atmospheric probe; it did not resemble a manned craft from the shape and movement of it. The orange ball and the white oval, I believe were a spherical, and a saucer like craft. They were one-shot sightings, but the rest of my sightings appeared to be the same type of craft. It was the diamond, the gigantic one, the small one, and the medium one.

My drawings and model of the diamond, I showed to someone who worked around the Tocomac fusion reactor in Princeton New Jersey, the same type of reactor that was on Beyond 2000 in Japan. He was told that a diamond is the most efficient shape for a fusion reactor. I now believe this diamond shaped craft is a fusion reactor. I have thought so for a long time, because I've been researching the whole time I had sightings. What I think is that, somebody has a fusion antigravity-ship; I've seen it many times.

It's definitely an antigravity ship; I also think that whoever made it doesn't want anyone to know what it really is. That's why it always comes out on pitch-black nights, where you can't see the true shape of the body. You only see the light pattern it will show you at that time. After all if someone admitted they have a flying fusion reactor, then they will have to admit that they have harnessed the greatest power source we've ever had, a power source that will end all energy problems, and which will make oil obsolete, and which will also make power companies obsolete. We should have had fusion by now in my opinion, but this and many other technologies have been suppressed, or should I say hoarded and kept secret, by whom? By the elite money controllers of the world, who operate in secret societies and secret government operations, and private corporate black projects, which are untraceable and not in the government realm. These people control trillions of dollars and have scientific knowledge that has been kept secret for centuries.

In my opinion this great fusion reactor is one of the many ideas stolen from Nikola Tesla. In my opinion he was the greatest inventor in human history. He was an electrical engineer who had close to 2000 Patents. He was born in 1856, and died in 1943 and the government confiscated his laboratory and everything in it, which was in Colorado Springs at the time. He was the father of wireless transmission of power, which the powers that be claim can't work, but it works, all of his inventions worked. They worked the first time with no working models. He conceived his inventions in his mind and put them together and they worked perfectly every time, just like the Niagara Falls A C. generation system.

I heard it stated in a documentary about Tesla, the Tesla coil was designed to lead to a practical method of harnessing fusion power, and that Tesla believed that ball lightning, or contained plasma could be the starter and the container for a fusion reaction. The ribs going down the sides of these reactors are Tesla coils. The squares are superconductors, the craft has its own gravitational field, a person with a lot of scientific knowledge will figure out what it is after seeing it if they're lucky enough to get the right information in their research. This is one of the reasons the craft hides its true shape, and comes out on black nights, when you only see the lights it shows.

Tesla demonstrated a wireless remote control submarine boat at the world's Fair in 1899, while everyone was in the horse and buggy era. He's the real inventor of radio. His inventions led to L.A.S.E.R. and M.A.S.E.R., robotics, beam weapons, weather control, mind control, force fields, electromagnetically induced invisibility, antigravity, he discovered scalar waves, zero point energy, to name a few. I think he was the real father of modern UFOs. Countless Tesla patents have been stolen and used secretly by the big money wielding secret societies and their private corporations.

I absorb all of the information from the different researchers, who appear on Coast to Coast AM I can sniff out something untrue from miles away, and I know which researchers are talking fact, and which of them mixes both. I've been listening to Coast to Coast

AM since 2001. Before then in my research, I read about Nikola Tesla, I also read about Viktor Shauberger, an Austrian physicist who conceived of implosion technology, he studied tornadoes, and created water vortex turbines. He created designs for flying saucers in Nazi Germany, which were based on liquid vortex generators, with water inside. He said water could solve all our energy needs.

Other people have demonstrated electromagnetic antigravity, such as John R. Searle, who built a small flying saucer. It flew straight up and disconnected itself from its starting power source, and rose up and out of sight, antigravity works. There are many other brilliant scientists whose knowledge has been stolen over the past few centuries.

I read an incredible book, Genesis by W. A. Harbinson, I think it's the Rosetta Stone of UFO research. I also got a book called Manmade UFO's1944 to1994, 50 Years of Suppression by David Hatcher Childress and Renato Vesco. It's a hard to get book put out by David's own publishing company.

Years ago my cousin and I researched the sources for Harbinson's book, and we found out a lot more information. We checked out a book called German Secret Weapons of World War Two by Rudolf Lusar that showed the remote-control flying disc, created by Flugkaptain Rudolph Shriever and also a "flying top". I think the flying top somewhat resembles the craft that were down in Gulf breeze Florida. I also read that Victor Shauberger was brought to America after World War Two, and he worked on secret government projects until his death, and that he kept repeating on his deathbed "They've taken everything from me, I don't even own myself."

More strong opinions from research

I don't want to get into all my research here. But I will say that in my opinion 99% of all UFO's seen are from Earth, I leave open that 1 percent for what I think are a few extra terrestrial races who qualify as fallen angels, because, as the Bible says they forsook their proper dwelling place in the heavens and came to Earth against God's will, as part of a great rebellion in the heaven's. Or if you consider it in the language of the sci-fi show Star Trek they violated the prime directive, not to interfere with a civilization's growth.

I think that civilization has been interfered with since the rebellion began, and that a time for a judgment is coming soon. I believe that our energy problems were solved by Tesla a hundred years ago, but his inventions were suppressed, because of the rich controllers of oil and energy in general. Free energy and antigravity should have been ours a long time ago. When the evil rulers of this world are cast out, maybe man will have this technology to help clean the Earth up. We have been 'dumbed down' and stolen from by secret societies, some of which are hoarding technology that goes back thousands of years. I think there is an ancient connection, but the most amazing advances have come about since Tesla and the advent of computers.

I have also heard about 'Projekt Saucer' which was a secret work carried out in Nazi Germany. Among the secret weapons they were working on at the end of the war, such as the V 2 rockets and vortex gas cannons, were the remote-control Feuerball, which I mentioned earlier, from Vesco's article and what he called the first flying saucer, the Kugelblitz or ball lightning automatic fighter. Projekt Saucer is said to have produced five different types of saucer, the first two being jet propelled with suction and expulsion of air, and the Flugelrad or wing wheel which was somewhat like an internal helicopter blade. The last two projects are said to have been electromagnetic gravity reversal technology.

The same technology was described in the early '70s by a UFO abductee named Carl Anderson who worked at Long Beach California with the Navy. He was on a UFO documentary, years ago in the '70s on Channel 9. I have that documentary on video, along with most every U F O documentary made since. Space people told Mr. Anderson their saucer was essentially a flying electrical motor that reversed the Earth's gravity when it reached the rotational speed of the Earth on its axis. Reverse polarity made it come back down, neutralizing the effect made it hover, he didn't explain the horizontal flight technology in that interview but there are many ways to achieve it. The Earth is one big electromagnetic generator. You don't need to be a super advanced alien to do this. In fact Tesla went beyond that almost a hundred years ago.

At one time Tesla had a cigar shaped craft floating around his laboratory called a Telautomaton, he said it was a reaction craft propelled by the "total reaction of sound waves". It's interesting to note that the very first modern UFO sightings occurred in the 1890's and were of cigar shaped craft.

I would like to point out that in the 1975 documentary 'UFOs Are Real' there was a portion of the film where the narrator was speaking theoretically and he said that we could build a nuclear fusion powered starship which could travel 37 light years in two months pilot time, and that it could go just under the speed of light, and when it returned the Earth will have aged 37 years, and the pilots will have only aged two months. He also said that Einstein's theory of relativity has been experimentally verified.

I can't help but think that the narrator was giving us a sneak peek into secret technology. Years ago, in the Cash-Landrum UFO case, two women and a child were in a car driving when they saw a diamond shaped craft escorted by military helicopters, they thought the craft was malfunctioning and was going to crash. They received severe radiation burns as a result of being too close to this craft. I think this was part of the early experiments with diamond shaped craft, and this experiment did not fare well.

Ancient technology used in modern times

Here's another huge point, I think the Coral Castle in Florida was built with the same technology that built the pyramids and the other gigantic stone megaliths around the world. I hear that Edward Leedscalnin, the man who built it, consulted with Tesla to build one of Tesla's gravitational field conversion motors which produces antigravity effects through electrical currents which can float the stones like balloons, by running the current through the magnetic poles of the stones, and, can also produce a plasma laser like cutting device to cut the stones.

I think this is part of the ancient technology which man was supposed to have a long time ago, which leads to free energy, antigravity, and a laser like device. Because of my mechanical theoretical mind, I relate to Nikola Tesla very much. I believe he was a very good man who wanted to solve the world's energy problems, but he was way too advanced, so the control freaks stole his inventions, just like they've stolen the inventions of countless other scientists probably over centuries.

One day when just rulership returns we will have these technologies to help clean up the planet and to get rid of polluting sources of energy once and for all. If you saw what I've seen you would do some heavy research too, but in my case I've been lucky to be pointed towards some amazing information from different sources, and I have an incredible memory. I remember everything I've researched in the last 20 years.

Someone like me belongs on Coast to Coast AM. My research has satisfied me and I don't want any trouble with the powers that be. I just want the ships to leave me alone, and to live my life and gain more freedom.

The reason that I thought that the sky quake may have been the big ship exploding, was because, if a big fusion reactor let go somehow, it would vaporize itself in the explosion, and the power might come out like plasma in all directions. Maybe these fusion reactors are too dangerous for now; if that's the case maybe we should leave them in space and beam down the energy they create, that's one of my many ideas.

The UFO landing in North Hudson Park

Here's an amazing case that I found out things about which no one has yet discovered. In the book Missing Time by Bud Hopkins, he mentioned a case that happened in New Jersey in 1975. There was a landing by a saucer type craft in North Hudson Park that borders North Bergen and Fairview. This is the park near the area where I had my first sighting, it is a county park, and it was called North Hudson Park back then. Now it is called James J. Braddock Park, in memorial of the local boxer. This is where Hudson County and Bergen County meet. In the previous weeks before the landing, this same craft had apparently scouted out the area and flew over Fairview towards the Park.

This craft had stopped by an apartment building and hovered for a few minutes close to a window in view of a family living there. They were the Wamsley family. I knew them. I went to school with their son Robert, in grade school. He had told us about his sighting, I used to hang out with him and ride bicycles, I remember fixing his bike when he had a flat. I heard from him and from other people how he got on television because of that, along with his parents.

They said it hovered while tilted in the air, almost like it was looking in the window at them. Then it slowly flew away, and they went outside to watch it fly toward the park. They described it as a double convex shape with the white light coming from windows along the sides. They heard no sound. The next weekend it came back and landed in North Hudson Park on one end of the baseball field. It was witnessed at around 3:00 a.m. by a storeowner returning from work in the city, and by a security guard in a nearby building, and by another eyewitness not far from the area.

I had heard about this landing many times when I was young, because I was raised in that area, I lived in North Bergen in the '60s and '70s, then nearby in Fairview from the '70s into the 80's. The closest eyewitness to the landing reported that the craft had come down and hovered a few feet off the ground over the ball field, and a ramp came down and a group of small beings in one-piece outfits, who he described as looking like kids in snowsuits, seemed to take soil samples from the ball field with scoop like instruments.

Something amazing happened after this landing that was part of my childhood. Because I used to ride bicycles and explore in that park with my friends in 1976 and 77, and we heard about that landing a few times from different people. There were more eyewitnesses to that event that Budd Hopkins never spoke to. It was a tight community where word got around fast, and it did.

The mysterious draining of the lake

The amazing thing that happened, was something that I realized and pieced together 13 years later, after reading Missing Time by Budd Hopkins and getting the details about the landing that I did not have, and got more information from another book, as well as my own memories of something that happened back then. Not long after the landing they drained the lake that I used to ride my bicycle around. The landing happened in 1975. A few months later a large excavation project was begun at the small lake bordering the ball field where the craft landed. This was an ancient duck lake where ducks and geese stop on their migration routes. It was about three-quarters of a mile to walk around this lake. I rode my bicycle around that lake more times than I could count.

This massive project was started which lasted a year and a half where they drained the lake and let the water flow in wide hoses down the hill by the edge of the park, down to Edgewater and into the Hudson River. After draining the lake a massive excavation project began and they dug out the bottom of this lake. This lake was only about 10 ft. deep, and it had centuries' worth of duck and geese droppings built up on the bottom, it should have gone to some factory to make fertilizer, processing it. They removed something like 6 ft. of bird droppings at the bottom of the lake and removed it in large trucks with large excavation equipment.

Now when I was a boy riding my bicycle in the area, I rode my bike on a section of the dry dirt of the lake bottom, after the soft dirt had been removed and there was harder soil that had been pushed around on the surface. That lake was left without water in it for about half a year, this whole project took about 18 months.

For me it was an amazing novelty, because I had never seen that lake empty, I had been raised in the area all my life up until then, I had been playing around that lake all my life it seemed from time to time, and I had row boated on it many times. The rowboat rides on the lake were a very fun thing for the community, a great thing for parents to do with their children; I loved it as a small child. Later they got newer pedal boats and continued the boating on the lake. That was a natural duck lake that had always been there, and the boat rides had a long history of fun for the community. I wondered when they would refill the lake, and why this project was taking so long, and why it was being done. I asked a lot of people but no one seemed to know why this project was being done.

They excavated down to a certain height below the bottom of the lake while loading the soil into trucks and trucking it out to who knows where. They covered the ball field where the craft landed in a layer of soil, and trucked away the rest of it. After they had removed a tremendous amount of soil from the bottom of the lake, and the lake being about 7 ft. deeper on average, they cleaned it all up and fixed up the outer edges of the lake where the walk way is, and they refilled the lake, pumping river water back up into it, finished cleaning up, and the project was over. I could never figure out why this project was being done, and no one else knew why. The project was finished some time in 1977, two years after the landing. This whole episode disappeared into my memory. Now I'm going to relate it all to the UFO.

Like I said, it was about 13 years later when I read a copy of Missing Time by Bud Hopkins and I read about my friends the Wamsly's and we found out the details of the UFO landing. By now I had seen the big ship, and was doing my own thinking and research. Here is the amazing thing; it's what I pieced together from the events of long ago. That was most likely a secret government mining operation to extract minerals for use by the people who make these UFOs; I'm talking about that whole excavation project under the lake. Here's why.

An element that defies gravity

I read another book dealing with UFOs and other unexplained subjects. There was a section of the book where a scientist was explaining some odd peculiarities with the elements that had been discovered. Among these was, that there was a particular type of basalt called lintz basalt, which is blue and is found in deposits separately in areas with a lot of basalt. The Palisades are made of basalt. The Palisades are the cliffs that make up that area where the park is. They make up all those cliffs along the Hudson River in New Jersey. The scientist in this book was saying that tests were conducted on lintz basalt, and that it was found that lintz basalt does not conform to the laws of gravity. It was found that lintz basalt will not accelerate in freefall. It does not defy the law of gravity, but it does not conform to it either.

This element has unique properties to it that seem to be magnetic, and interact with the Earth's magnetic field. It is found as blue clay. Now I remembered some things from the past that really tied this whole thing together. There was a layer of blue clay beneath the duck droppings on the bottom of the lake, which was 15feet thick at one end of the lake. That layer was underlying most of the lake bottom, and all of it was trucked out, and none of it was returned, unlike much of the regular soil that was brought back and layered over the top of the ball field where the craft landed.

That thick layer of blue clay is what they were after. Back when this project was going on, someone I know spoke to the supervisor of the heavy equipment operators who worked for the excavation company who was doing this project, and he didn't even know why this job was being done. The company who did the job went bankrupt. It was a huge project; the cost had to be gigantic. There was something really strange about that job back when it was happening; no one knew why it was being done. They got hundreds of tons of that blue clay.

I think this was a secret mining operation to retrieve this material. This material sounds like the perfect thing to add to alloys that would be used in making antigravity craft. Seeing that this material already does not conform to the laws of gravity it would be perfect for the job, and who knows how it interacts with electric currents. I remembered the fact that when the craft landed, they took soil samples, and that in less than a year after a major excavation project was under way.

I believe that UFOs are capable of advanced spectrum analysis of the Earth below them, and can find deposits of minerals and metals and other elements, and that they knew where this material was concentrated. This was a county park. This seems to be an example of state and county government working together and using an excavation company

84

to carry out a mining operation under the guise of something else, to be used by someone else. I believe this was an example of a secret government mining project.

There is no other logical explanation as to why this job happened. The people who did the job didn't know why it was being done. And my father knew it back then. Many researchers believe that secret mining operations are being conducted all around the Earth, at the poles, in other remote areas including on the moon that is made of light metals. Light metals make alloys, and I will remind you that the Shawangunk Mountains have light metals in them as well as quartz crystals. Also that hum that I heard at night seemed to be coming from underground and may have been tunneling machines under the mountains. Our house was sitting above shale and picked up the vibrations of this sound on many occasions because the rock mass under us was picking up the vibrations. Apparently the saucer landing led to the excavation project.

This whole episode really blew me away because this park was my childhood romping grounds. The memory of that lake being empty, and me playing in the dry lake bed, and trying to ride BMX style bicycle in it, is a very sharp memory for me. Also it's incredible to me that UFOs were in my childhood area, and that they were in the area that I moved to as an adult.

It made me think that I'm caught up in a cosmic freak show that followed me around. Real life is more entertaining than fiction. I have gotten one hell of a show. I wish I had all my sightings on film. It would be in incredible film.

Now I'm thinking that this lintz basalt is loaded with magnetite because the ducks and geese use forces of magnetism to direct their migration, and this was an ancient lake, not manmade, it was in modern times that they added a cobblestone walkway around the edges of the lake, and later poured sidewalks over it. I had no idea my UFO research would carry me all the way back to 1975 in my childhood area. I knew my friend Robert Wamsley was completely believable when he described his close encounter. I was 12 years old when that landing occurred in 1975, I was in my bicycle exploration stage. I would think that something filled with magnetite would be part of a propulsion system in a machine that reverses Earth's gravity. I would think that if you concentrate magnetite after refining it, you could use it in a propulsion system running electric current through it and rotating it in a vortex generator or a large electrical motor.

By the way, the man who drove by at 3:00 a.m. and was the closest eyewitness to the landing, said that he heard a sound from the craft which was like a refrigerator humming, and that it slowly rose up to about 150ft. in a 45 degree angle, and then shot away at incredible speed which greatly accelerated, and it shot away over the horizon.

Another eyewitness to the landing, a security guard in a round building called the Stonehenge apartment building, which is directly opposite the baseball field where the craft landed, watched this thing land and was scared and went to pick up the telephone to call a tenant, but the second he picked up the phone, he said he heard a high pitched vibration and saw a crack on the window of the lobby a few feet from where he stood, and this intimidated him and made him put the phone receiver back down. Later it was found that a hole had been gouged in that glass which was quite thick. A marble sized piece of glass had been removed. I think this could have been an ultrasound beam weapon, if not a laser. More details of this landing are in Bud Hopkins book Missing Time.

I have given you true testimony and detailed descriptions, now I'll give you more opinions from my research.

UFO's and Ancient Scripture
My opinions in a Biblical standpoint

Some people think if UFO's landed and were made public and the whole world knew it, that it would deal a tremendous blow to the religions of the world. That view is created by people who are anti-religion to begin with. Or, who are ignorant of ancient texts and what they say about the subject. I think that view is completely wrong. I think this is destined to happen, and it will prove the Bible correct, like I said, I think most UFO's are from Earth, however ancient history shows us that some of them are the product of fallen angels. The Anunaki were apparently sons of the fallen angels, who were giants, who deceived the Sumerians and lied to them, telling them that they created the human race.

I believe the so-called aliens will return and they will lie like they've done in the past and tell us that they are God, and they created us, and now they're here to prevent us from wiping each other out in a nuclear war, and that they will bring us peace. The Bible refers to the great deception, and the final world government under Satan, the great lie by the great deceiver, to turn everyone away from the real God, who is love, and doesn't need technology. Technology is the golden calf that the world worships. I think that researchers like Stephen Quayle, Patrick Heron and Lynn Marzulli are right on the money with their research, when they say that the final deception will be brought on by UFOs coming down, quite possibly, after a limited nuclear war, and pretending to be our saviors.

I think John Lear honestly told us what he was told, but I think some of what he was told was false, controlled information. Once again in his case the aliens tell us basically that they are God, that they created us, and that they created all the great leaders of religions. This is part of the great lie. I think when the alien's come down they will fulfill Bible prophecy by showing themselves to be the great deceivers. Satan presents himself as an angel of light bringing enlightenment.

The secular world is completely ignorant of all spiritual subjects and is completely vulnerable to the attacks of lying entities from the spiritual and physical realms. UFO abductions and other experiences show that they are hostile, they rape people and experiment on them, and they deceive and lie regularly. They kill people and animals. They are part of the great rebellion against God in the heavens, and the ruler of the demons is a super scientist. These so-called aliens cover their ears and run away when they hear the name Jesus Christ, this is no wonder because it has been only in his name that demons have been cast out, no one has ever cast one out in another name because his authority is the ultimate one in the spiritual realm, and all power has been given to him in heaven.

For now, until the great judgment, we have demonic enemies with great power through their technologies, many of which they have stolen from us. I think many UFO's belong to our government-corporate- private sectors on Earth and to secret societies and their private corporate realms. There seems to be a war going on by UFOs in our atmosphere, and around the Earth. I think different factions with UFOs on Earth are at war with each other. Also some factions with UFOs on the Earth may be at war with UFOs that are from the demonic fallen angels who left their proper dwelling place in the heavens to interfere with and exploit mankind to our detriment.

Those who say the so-called aliens are benevolent space brothers are deceived by their lies. A good case study of UFO abductees and other encounters with UFO beings clearly shows they are hostile and manipulative, they are masters of deceit, they have committed mass murder on humans and animals, in one case killing a whole village in Africa. Just like the episode of The Twilight Zone called "To Serve Man", if you believe their lies and want to follow them you are at risk. Only 1% of UFOs could be extraterrestrial. This phenomenon is mostly from Earth.

Harbinson's book "Genesis" says that the greys are astronautic cyborgs created by scientists in a secret society in Antarctica which is a masters and slave's society which grew out of Nazi Germany, a godless society devoted to science. To answer the real questions of UFOs one will have to be drawn into the spiritual realm, because it is spirit forces, good and evil, which guide this phenomenon to its ultimate end, though there are physical forces on Earth guiding the construction of the machines.

The Nazi scientists were guided by demonic power whether they knew it or not, they built flying saucers, and they dissected people in their death camps. They wanted to see how much G- force a human body could take. They found that an eight year-old child could handle five times the G force as an adult. The original close encounters in the '50s had beings that looked like bald Oriental children in tight fitting one-piece garments; there were no greys back then. Harbinson says the Nazis took children and robotized them into the pilots for their craft, that they were Ache Indians from South America who looked Mongoloid or somewhat Oriental. It has been written that Harbinson was former British intelligence, I don't know if this is true, but his book is the most important book written on the subject, and the only one written which points towards the Earth based origin of UFOs. I have seen tremendous amounts of evidence that UFOs have been made here on Earth. I see no solid evidence indicating a strong case that they are extraterrestrial.

I think there's a strong case for ancient extraterrestrials who were fallen angels. Giants were in the Earth in biblical times and they were crossbred from fallen angels. Giant skeletons have been found all over the Earth on every continent; their remains have been found and have been scouted out by people like Robert Ghost Wolf and Stephen Quayle. Mainstream science goes out of its way to hide the giants from the public. The Romans were killing giants in biblical times, so was King David, and so were the Native Americans at roughly the same time. At that time the fallen angels came to Earth and the giants spread out to every continent, and were ravenous and murderous and everyone hated them and fought against them. The American Indians killed them and put their bones into burial mounds and caves. A large number of them were cannibals.

Bible Accuracy

The UFOs will return with their great deception, and for those who know their history This will be proof of the accuracy of the Bible, just like the giants are proof, and the secular world tries to hide that proof the best they can, but I applaud researchers like Robert Ghost Wolf and Stephen Quayle, who dig up the truth about giants. For that matter anything that proves the Bible correct is hidden by the secular world. And there is a tremendous amount of scientific evidence proving the Bible correct, contrary to the lies and spin created by the media. Hundreds of arguments trying to prove the Bible incorrect were all proven to be false arguments put together by taking statements out of context and contriving ideas, there is a subject for another book, another gigantic cover up just like the UFO cover-up.

Ancient texts from the Dead Sea Scrolls tell us that UFOs and giants came down together after the great flood at Mount Hermon in Israel, 200 of them. The previous giants were killed by the great flood. Contrary to popular lies put forth recently in the media the accurate ancient texts do not contradict the Bible in any area; in fact they give much information that proves the Bible texts correct. The Dead Sea Scrolls include fragments of every book of the Old Testament except one, and proving they haven't been changed at all and 5000 years, except for 23 letters that were later corrected, and I mean individual letters as in the alphabet. The Chester Beatty papyrus proves that the New Testament hasn't been changed at all either, contrary to more popular lies in the media lately. This papyrus came from 200A.D. and was one of the very first copies of the original text of the Gospels. Master historian Sir Frederick Kenyon wrote seven volumes on his study of the Chester Beatty papyrus, and proved that no substantial additions or subtractions were made to the New Testament from its writing until now. Today the media is obsessed with this lie that the Bible is edited. This has been proven false.

The church had nothing to do with the writing of the Bible. It was formed by first century Christians who had nothing to do with any church, or any other power structure on Earth, their kingdom was not from this Earth, as Jesus said. Other ancient texts add more information to the accuracy of the Bible, and some of them discuss UFOs and fallen angels. The giants came from the fallen angels, and were evil and murderous, and called themselves gods and wanted to be worshipped, while they murdered, raped, and bullied everyone in their rebellion against the creator God.

The Bible and other ancient texts need to be discussed in this subject, because that's when UFOs first came down, with the rebellion from the heavens initially, then with the coming down after the great flood at

Mount Hermon in Israel. Accurate Extra Biblical texts such as the book of Enoch add more information to what's in the Bible about the fallen angels. There are no historically accurate ancient texts that contradict anything in the Bible, they all support and add more information around it. The Gnostic texts are not authoritative, they are philosophies and opinions of different and often contradicting philosophers influenced by Plato and Oriental mysticism, the Gnostics were self styled and following their own ideas, seeking their own followers and contradicting the scriptures of their day, many were immersed in old pagan ideas. The post Christian Gnostic, Marcion was known to be a liar and manipulator who chopped up and edited the gospels to his own ends. The early church fathers called him the "spawn of Satan".

Post Christian Gnostics made up their own texts mentioning Jesus or the apostles, beginning from the earliest; 60 to 100 years after the original New Testament documents were completed, and after all of the eyewitnesses to the miracles of the apostles and Christ had died. These texts were falsely attributed to apostles by using their names in the titles. They were rewriting history to their own ends just like the new world order deceivers, and Illuminati Luciferians who control the media and lie about the Bible today. These people had no connection to any apostle or authoritative figure. Many different ideas and agendas are reflected in those texts, and many of these ideas conflict with each other.

The Nag Hamadi library had a collection of Gnostic texts copied into Coptic starting 200years later because they were doing a study on heresies of the church, which had been rejected as fantasies, and the creation of people's imaginations, philosophies, and agendas. Post-Christian Gnostic texts are ramblings of ideas and not real history; they make no pretense of being real history. It's only today that some people try to pretend that they are real history because they are trying to rewrite history. They are simply written texts and not Gospels of any kind. The use of the word gospel does not apply to any other texts besides the first four books of the New Testament.
Prophecy proves what comes from God and what does not.

The word "gospel" means the greatest good news we can receive. This good news was the conquering of death, the overriding of sin, and the reestablishment of eternal life for obedient mankind through Christ's sacrifice.

Don't believe any of the conflicting "Jesus was a regular guy" stories that abound in spurious texts. There are many conflicting stories, which should remind one of the many conflicting false witnesses who were called upon to testify against Christ by liars seeking false testimony against him.

The anti Christ world government is coming together with the help of media brainwash to rewrite history and to knock down the faith of Christians with lies about the Bible. They also want to get us all to follow the new-age saviors who will come in UFOs.

A perfect example of media spin and brainwash is the book the Da Vinci code by Dan Brown. This book slams you with an apostate message in the 55th chapter. It is a compilation of lies told against Christianity. Because people like him hate the church, they need to lie about the Bible and rewrite history. The Bible was not written by the church. Neither was it changed by the church. There are at least 70 separate lies and distortions in the Dan Brown code. There are no real facts in it. I think there's no excuse for slandering a religion like that, with all lies and spin. Why doesn't he try that with Islam?, that phony.

Nowadays it's only okay to rip apart Christianity with lies because of the anti Christ world media. The early Roman church mixed paganism with Christianity that produced Christendom, a false polluted Christianity. Real Christianity was based strictly on the mission of Christ and the apostles, and on the testimony that came from the hands of the apostles, who were living first century eyewitnesses to the miracles of Christ and the miraculous workings of the Holy Spirit upon themselves and the people they ministered to.

People ignore the real issue, which is the undoing of death and the resurrections and healings being proof of the real messiah. People should check out some of the codes that are truly interesting like the Parthenon Code as explained by Robert Bowie Johnson. The Coast to Coast AM round-table discussion on the subject of the Da Vinci Code was a disgrace, and did more to bolster the lies, with no one there to defend real history. This was one of the very few shows they had that was a disgrace.

Most of the time Coast to Coast AM is doing us a great service with great researchers, but once in awhile you have people who tell some truth and some lies, and other times you have people who are just liars, but that is extremely rare. Nowadays I'm a big fan of Coast to Coast AM because it's the only show around dealing with paranormal subjects, and it is a great show over all, with great hosts, doing us a big service, thank God for Short Wave radio, and some AM, the last form of media not taken over by the spin doctors.

The last days of deception

We are living in a day and age of politically correct brainwash, where no one can call a spade a spade, and no one can tell truth from lies, or call them what they are. We're living in a day where it's ok to bash Christianity with lies and it's encouraged and continually done by the media, but if you dare to say a negative word about any other religion then you are called intolerant, it's anti Christian brainwash. They have tolerance for everything except Christianity, including terrorists, because they are under a demonic spirit. It's called cowardice and absence of morals and historical knowledge, combined with an evil agenda to steer us away from salvation, and towards slavery and death.

Politically correct is a phrase which came from Stalinist Russia and is actually a brainwash to remove the spirit of truth from people and turn them into moral weaklings with no moral center so they can be molded into a rotten new styled communism which is anti God, anti-Christ, anti humanitarian and pro mass murder. Five corporations control all the media, they are controlled by secret societies who are Luciferian and dedicated to the new world order which is the final satanic world government the Bible warned about in Revelation.

Christianity is the only religion being attacked nonstop by the media, because the controllers of the media are devil worshipers. The secret societies that control media corporations are part of the globalists including the 14 richest families in the world. The best researchers on these subjects in the world agree that they are Luciferian conspirators under demonic power. Jim Mars, Alex Jones, and Allan Watt, to name a few will agree, and they have examined their writings for decades. They already have all the money in the world, but it's not enough, they are control freaks who want human souls.

Jesus is the only one who can cast out demonic power. I have heard five different exorcists from five different religions interviewed on Coast to Coast AM, they all said that demons can be cast out, only in the name of Jesus.

I have examined Native American prophecies by Speaking Wind, whose tribe is a cousin of the Hopi's, his prophecies about the upcoming Earth changes perfectly lined up with the book of Revelation. He mentioned the great earthquake, the darkening of the sun, the changing of the landscape, the coming age of peace that they call the fifth world, which is the same thing as a thousand year reign of Christ. Where peace will be brought back to the Earth after the environment is set straight, then time itself will cease, along with aging and death. Then there will be a resurrection of all who obey the true God of love. This is the return of Eden or the fifth world, and the big cleansing will be complete by the end of 2012 according to Speaking Wind.

Jesus said the whole world is lying in the power of the wicked one. He also said the ruler of this world will be cast out. If the power structure of Earth were not in the hands of demonic people, we wouldn't be slaves to obsolete energy forms, and we wouldn't be killing each other in wars that are created by globalists, and their profiteering corporations. The ruler of this world only attacks Christ, his nemesis, because he is the enemy of God and mankind, and is working to destroy us. Those who follow his spirit are putting together the final world government that will likely be controlled by the U.N., and they will do everything in their power to destroy Christianity, and murder Christians. Today the United Nations is pro radical Islam. In the last decade the U.N. has presided over the murder of millions of Christians in African countries by radical Islamic terrorists.

Now the new world order puppets are enacting hate crime legislation that targets only whites and Christians. If we don't stop the U.N. and the globalists, and their communist friends, and the other evil forces in the world that have penetrated America and are currently destroying our country, we will have tyrants worse than Hitler enslaving and murdering mankind. If people will wake up from their dream world of entertainment, which has been implemented to dumb them down while their nation is stolen from them, they can help turn things around with the power of prayer and action and education for human freedom exposing what's really going on.

Either way in the end God will win and the prophecies will be fulfilled, and every evil soul will be banished from the planet and the new era of love will begin with Christ as king because he conquered death and showed more love than anyone, and made the sacrifice to redeem mankind. He gave us an example of love to follow which involved healing everyone and thanking our creator who is love, and who gave us everything perfect before the great enemy came and messed things up.

This enemy will be cast out soon. I expect by 2013 we will be in a new world after a great cleansing and destruction come. People who refuse to believe what's in the Bible never studied it. Its prophecies have been a hundred percent accurate, 2000 of them. Check in with PHD Chuck Missler, a brilliant Bible scientist who has some amazing statistics and info regarding the Bible.

The Mayans talked about the end of time, at a certain point we are supposed to realign with the galactic center, then time will cease. Time seems to be a byproduct of the Earth being off kilter, tilted the wrong way on its axis, and not properly aligned with the galactic center. This will probably change after 2013, and we will enter the new era of peace. The thousand year reign of Christ is the same as what the Native Americans call the fifth world. This is the age of the resurrection where there is no time, nor is there sickness, aging or death, it is a return to Eden. The Bible says it, and so does Speaking Wind.

The most accurate prophecy system in human history is found in the Bible, with 100 percent accurate fulfillment, I think Speaking Wind's prophecies are second to the Bible in accuracy, I have two books he wrote, The Message and another called When Spirits Touch the Red Path. It was in The Message that his prophecies were made concerning the coming Earth changes and his own death.

Undo the final deception with knowledge

A close study of UFOs will eventually bring you to the conflict between good and evil and into the ancient scriptures because this is where the phenomenon started. But I believe most of the machinery being seen in our skies today is made with Earth technology by governments and secret societies. I leave open that 1 percent for craft that may belong to fallen angels who are demonic and part of the great rebellion. I think that if there are any good UFOs they are the ones that come from the secret Marconi base, because they were students of Tesla and were not involved in any secret societies related to the globalists and the ancient manipulators. They were the ones who wanted clean Earth friendly power and harmonious existence.

As for the idea that angels may have UFOs too, I don't believe it, they don't need technology, and they are not part of it. Technology is the product of the fallen realm, it's only the fallen angels who descended into the physical realm who would do these things, using technology for manipulative evil ends, and interfere with us against God's will. Technology is deadly without spirituality, and the ancients will tell you mankind has taken this wrong path.

Free energy which does not pollute is all around us, from solar to wind to hydrothermal, and water power, steam power, wood burners, up to high-tech inventions like Tesla's gravitational field electromagnetic conversion motors, which is the best technology of all in my book, to his many other inventions based on sound waves, scalar waves, wireless high-frequency electricity.

I think Tesla created many forms of free energy, so have other scientists who have been stolen from. I have read that Tesla took a 1930 Pierce Arrow automobile and replaced its gasoline engine with an 80 horsepower alternating current electrical motor and then hooked up a black box of electronic tuned circuits which were adjusted to the frequency of the Earth, and that he ran the car off of the Earth's magnetosphere, the ultimate free energy motor, they drove the car around for nine days up to 90 mi. per hour, free energy no fuel, it ran off of the Earth's power because the Earth is a giant electromagnetic motor just like all the other planets. We should be working on recovering the technologies that have been stolen from us and from Tesla.

Nothing but updated deception from media

UFO sightings en-masse began in the forties. For 30 years we were lied to and told that we were seeing things. Then, as of the '70s, all of a sudden the media was telling us that they're real and they're aliens from another world. The media has been shoving aliens down our throats ever since. No one ever talks about the possibility that they are from Earth.

The authorities of the world will never tell us the truth; their long history of lying says so. I have heard that the United Nations has a world constitution already written, it's called the Federation of Earth. I think that's the new one world government that will be brought together by a mass UFO landing and official contact with the governments of the world, the great deception.

I agree with Jacques Valle that they are messengers of deception. After reading about many abduction cases, it is clear that these so-called aliens abuse people and use mind control on them while conducting medical experiments on them against their will. They are no better than common rapists and murderers, they are Godless and emotionless, carrying out their scientific experiments in the same manner as Nazi scientists did. They cause pain and suffering while violating people's human rights, and then leave people with a posthypnotic suggestion which is friendly. By their

106

fruits you shall know them, they are evil, they are part of the demonic rebellion. The Bible says that Satan and his rebellious angels were cleansed out of the heavens, and banished to the vicinity of Earth where they await their final judgment and destruction.

The scriptures have said "woe to the Earth and sea, for the devil has come down with great anger knowing his time is short". The great liar deceiver who has made mankind suffer for thousands of years is about to be judged, along with his gang of dregs of the universe, and they want to take us down with them. I hope that people out there will check with the scriptures, and see through the coming great deception, these demonic creations torture and abuse people and animals, and they expect you to believe they are Gods. They are deceivers and butchers.

I'll bet on the good book

The real God is a God of love and his kingdom will come after these false gods have been cleansed off the Earth in the coming tribulation and Earth changes. The Bible says the God of heaven will set up a kingdom that will reign forever in love and righteousness, on a cleansed, peaceful Earth. The new Earth is coming after this cleanse. We don't need technology without spirituality; we need the removal of the evil power structure. Everyone can do their part, and we should all pray for God's help in getting rid of these demons and their power hungry puppets. We will need God's help, and he has promised to help. He has promised a new Earth free from evil and we will see it.

Let's work together to get through the cleansing. Working together and agriculture are the keys to surviving the cleansing, with prayer and God's help. The real God doesn't need UFOs or technology; they are the tools of the false gods. I hope you all see through the coming deception. And steer clear of the ships! That's part of my take on what I've seen and I have more facts, and more reasons for having such a take on things.

I can answer questions in a lecture format and would like to get on Coast to Coast AM, which is the only show I know of dedicated to paranormal subjects. I appreciate the existence of such a show. I say don't look to UFOs as your savior or your helper, they cannot save and they haven't helped us. It still pays to be observant and cautious in today's world.

It has been stated by many, that UFO type craft and related high-technology devices have been back engineered from crashed alien craft. After looking into hyper-science and many inventors of the past, it is my position that this statement is not true and never has been. I believe this is part of controlled disinformation. Extremely advanced technologies have been created by amazing inventors in the past, and have been suppressed by those who control global oil and power generation.

Enough patents have been stolen and suppressed to build antigravity craft, force field technology, subatomic particle beam technology, lasers, masers, weather control, mind control, electromagnetically induced invisibility, brain implants leading to telepathic communication, scalar wave weapons and free energy as well as scalar wave weather control, mind control, and God knows what else.

These inventions were stolen from Tesla and other scientists who wanted to help the human race solve problems; however they have been stolen by secret societies and the corporate and military elites, controlled by the real money wielders in private black operations. These technologies have been taken to their ultimate usage.

If many of these inventions were ever admitted to, they would also have to admit that we have limitless free energy, which is something the energy manipulators will never permit. Little pieces of UFO disclosure Information leak out, but from what I've heard they are mostly disinformation. I say the scientists most responsible for UFO technology are people like Tesla foremost, also Victor Shauberger and later experiments in antigravity were performed by people like John R. Searle and T. Townsend Brown. Earlier free energy devices may have been created by John Keely who created the musical sphere. Keely was a natural born subatomic physicist gifted with tremendous insight. His writings are hard to read and understand even for advanced scientists. He was able to discern energetic relationships between sound and light, and knew how to harness some of these energies into free energy devices.

We also have many conventional means of obtaining free energy that have not been developed and have even been suppressed or held back and slowed down. I look forward to the day when clean forms of energy dominate and the destruction of our environment ceases. This will come about by the hand of God's judgment, then we will be free to use these clean technologies which we should've had many years ago.

The secular world thinks that when UFO's land officially, that it will be the greatest event in human history. This is nonsense. The greatest event in human history was when the son of God came and resurrected the dead. And still greater will be the mass resurrection in the coming thousand year reign. When the great deceivers come down and cure many diseases, it will be the diseases they have caused by denying us of our own natural medicines and natural lifestyle. All the technology in the world cannot resurrect the dead. In 1500 B.C. we had a great war between the Gods, and the true God showed his face and his power and humiliated and destroyed all of the false gods in ancient Egypt. This episode is about to repeat itself on a larger scale with more advanced technologies including UFOs. This time the same God who created the universe will make his name known again and will destroy the false gods for good.

The indigenous peoples tell us that this has happened three times before when technological races became so big they spanned the entire planet, like in the days of Atlantis and Lemuria. The fallen angels came down back then and created a world wide technological power and went against the real God, and were destroyed. Those who go after technology are golden calf worshipers who follow the false god who is Satan. In those days God inundated entire continents. In the coming final war between good and evil the real God will destroy the UFOs no matter where they are, including underground, in the oceans, out in space, on the other nearby planetary bodies, and wherever else their rebellion is taking place. God will change the electromagnetic forces and all of their craft will be useless.

The Earth will be cleansed just like the heavens and there will be no more room for these technological murderers. They will be thrown into the abyss and the Prince of Peace will rule in a kingdom that will reign to times indefinite, and his authority will stop the rebellion for good, everyone who wants to follow the evil false gods will perish for good with them. Those who prove to be with the real God of love, after going through trials, will be the ones who live forever in the coming kingdom of righteousness, may you be among them!

A bonus updating chapter

Pay no attention to that machine behind the curtain

This updating chapter is being added in early May of 2007. Recently some photos of a strange craft were brought to Coast to Coast AM by someone named Chad. These photos are amazingly clear, and have brought about a lot of comments and speculation. I would like to give my take on this subject. I have received feedback from a high level computer expert who says that these images of this craft cannot be "pulled apart" and are not the product of "photo shopping" and appear to be real.

I have also received the testimony of someone I know who is a confirmed eyewitness of that large Hudson Valley UFO which appeared visible to the area between New Paltz and Newburgh on September 8, 1988, when this big ship was out and visible for a full 45 minutes. That night was extremely dark, pitch black would be a good description. That night it was impossible to see the true shape of the body, you could only see the lights, and the patterns of light that it chose to display. However, this eyewitness was one of the few people to view the ship from underneath, in the town of Newburgh, coming out of a deli close to Stewart Air Force base.

He had seen the incredible light show and went back into the deli and yelled out to the people inside to come outside and see this incredible object in the sky. A group of people came outside and witnessed this giant craft with its light show, no sound, and ability to tilt sideways in the sky while moving incredibly slowly. This giant craft with lights on it was at least a thousand feet long. This eyewitness is a very skilled observer who told me that the lights from the surrounding buildings and streetlamps partially illuminated a section under the craft which appeared to be solid, while the large lights on the craft appeared not to be solid, nor did they necessarily appear to be on something solid.

But here's the kicker. He told me that he is 99 percent sure that the part of the craft which was solid was exactly the same shape and structure of the craft which Chad photographed. He described it as a banjo shape years ago. When he saw the photos of this craft which were on the Coast to Coast AM website, he told me that was the solid core of the craft that night. That same night I had watched this ship from miles away on top of the Shawangunk Mountains from the parking lot at the hairpin turn. From that distance it appeared to be a giant craft with an invisible body in the darkness displaying lights in different patterns. The lights moved around freely going through three different light changes. The first was the same as the Pozuolli tape, a group of white lights in the front with a large red light trailing it.

Afterward was the arrow of red lights on the front bottom with the red one in back. After this came the boomerang of different round colored lights in a large V shape in the front.

My impression from that distance was that this was a huge diamond, because the different light changes appeared to fit onto a diamond shape. However I do agree that the lights were nothing solid, they were just projected light, and many of them moved around quickly and freely just like a computer display screen would operate. This brings me to my take.

Because of the amazing visual memory and testimony of this eyewitness, and after discussing it in depth with him and comparing what we both saw that evening, myself from a distance and he from underneath it with lighting around him, I've come to the following conclusion.

The craft that Chad photographed is real and I am now sure that it is a hologram projection machine which can create a large electromagnetic field around it which can then be used to computer modulate light shapes, sizes, colors, and patterns to display, in what ever movement it chooses. This technology has been discussed by physicist Fred Bell. Once you create a large electromagnetic field, there are then many different things you can do inside of that field. Another use for this machine which has been suggested to me is that it is for ground penetrating tomography, that the magnetic field it creates, with its electromagnetic radiation, is used to penetrate the Earth to see what's going on underneath the surface for internal mapping.

Both of these uses can be attributed to such a machine producing a large electromagnetic field. The electromagnetic radiation coming from it also explains the headaches that Chad was getting. I think it's obvious from the way Chad approached this, that he is not trying to fake anyone, that he's not seeking a publicity stunt because he has nothing to sell and has not tried to make money off of anything, and, because he was just asking for an explanation from an open minded audience.

I think it was kind of mean-spirited for some people to call it a hoax and question Chad and his motives. I think it's clear he has no motives other than to ask the most open minded audience in the world what they believe this thing might be, because he really wants to know. The evidence shows the photograph was not faked.

I want to get the opinion of physicist Fred Bell regarding this craft, because I'm sure he will know exactly what it is. For now I think someone is using ground penetrating technology to look into the Earth to enhance the abilities of H.A.A.R.P. technology to see into the Earth. I'm sure this machine has been used as a field generating hologram projector, computer modulated, and remote controlled from another craft. People have now pointed out smaller craft up above it in the photographs, which I noticed when I viewed the photographs also.

The 2 sightings of the thousand foot ship could have been the product of this hologram machine, because they were both on pitch black nights where the body was not visible from any distance, and the lights were large, colorful, blinking on and off, and moving around just like a hologram generator would produce.

The other craft that I saw were definitely not this craft. The small diamond was obviously exactly what I drew and made a model of. The orange sphere, the white oval and the black cigarette shaped crafts were all something completely different; however I do believe that the 2 sightings of the big ship could have been the product of holography. Like I said, let's talk to physicist Fred Bell, I believe he will know exactly what this machine is. New photographs and eyewitnesses of this craft have come forward in the ensuing weeks. People who say the photographs are hoaxed are absolutely wrong. They are not considering that there are many eyewitnesses in different places who have seen this craft, and photographed it. When a machine creates a large electromagnetic field around it, that field can be used as a projection screen for holograms. I think the "Chad" craft is remote-controlled by a smaller craft above it which can computer modulate and project light patterns onto the field created by the larger craft, but I think the main use of the craft is for ground penetrating tomography for mapping underground faults and caverns along with gas pockets, and it can probably see down deep enough to see magma.

I think it acts as a deep penetrating spectrum analyzer. After seeing the new photos in June of 07 showing the same type of craft with a bunch of mechanical additions, I would say there are different craft made which are similar, which perform different functions. It is interesting to note that these craft are floating around in an area that is extremely seismically active. I think the secret high-tech; black-op corporate government is looking deep into the Earth to map out where the fault lines really are. They can also find gas pockets of methane, as well as accurately seeing where the magma is close to the surface.

In my opinion this secret scientific government knows that the Earth changes are coming very soon, and these Earth changes are going to wipe out this secret world government of technology, they will try everything in their book of tricks to stave off these catastrophes but none of them will work. Judgment cannot be averted.

In my latest research I am reading about electromagnetic scalar wave technology which has been stolen from Tesla and developed in secret to create propulsion systems, free energy, weather control, mass mind control, and has been weaponized over the years by Russia and the U.S. in a secret arms race.

I am also currently reading about the propulsion and energy systems of ancient Atlantis that Edgar Cayce spoke about in his readings, and is now being written about by a NASA physicist. A new form of limitless energy has just been discovered by an American named John Kanzius who accidentally stumbled upon it. He found that he can ignite salt water using radio waves and it separates the hydrogen molecules and the salt water actually burns with a bright yellow flame burning at over 3000 degrees.

Once again the phony energy crisis has been solved, with an unlimited element called seawater. I'm sure the oil company thugs will find a way to suppress this. We can probably get 500 mi. to a gallon on seawater. This is an easy way to burn hydrogen and can be used in large steam engines to power electricity generation. This simple technology can be used to power automobiles and buses or trains. You can see the salt water burning in the original newscasts on Jeff Rense's web site where it says "man solves energy crisis" and "see salt water burn" video. Also, see "saltwater burns .com".

Colonel Tom Beardon has written about scalar wave electromagnetic technology. An amazing article he wrote can be read on the alternative news web site of Jeff Rense.

Another amazing article about Atlantean technology was written by NASA physicist John Sutton which can be found on the web at the Hutton Commentaries, Inc. where he discusses crystal gravity field generators powered by solar radiation, he put together some great research with intuition. I am grateful to read his research.

I am currently reading an article stating that Tesla was murdered after all of his super high-tech patents were stolen and weaponized by globalist thugs who handed them over to the Nazis to be developed and then, of course, took them back to give to their secret society puppet masters.

The Nazis were the first people with the metallurgy and scientific skills to put together the antigravity machines Tesla conceived, as well as many other inventions of his. The secret societies that backed the Nazis are still in power and have stolen the inventions of many scientists over many years. They are the technological masters of the new world order, and they play with people like their toys with their many cruelties like war and mind control. They are leading us toward World War three. They have stolen all of the free energy, clean technology, and healing technology for themselves so they can manipulate the rest of us. I'm sure they will cause much harm before they are stopped, but they will be judged, and very soon, by the same God who humiliated the false gods of ancient Egypt. The real God will judge them out of existence very soon, along with all of their demon puppet masters.

I encourage people to research and read about some of these amazing technologies that have been stolen from us on the Internet, you will see that there never has been an energy crisis, and that we have always had antigravity. There has only been a crisis in leadership because the nations have been hijacked by demonic power mongers, but not for much longer, they will soon face the fire of judgment.

About the author

A former tradesman, industrial diamond cutter, worker in masonry, roofing, industrial coatings, carpentry, landscaper, logger, and skilled musician, the author was born and raised in America, and has lived in the Hudson River Valley area his entire life, in New Jersey, then in upstate New York.

A man with many interests, hobbies, and skills, he is a skilled researcher and an avid reader who has read hundreds of books. Tremendous knowledge courses through his father and uncles. One of his uncles amassed 5 PhD's. He focuses his researching skills in whatever direction he chooses. Not much different from the average American, but he is a natural born scientist, and a good natural physicist. He has studied up on many amazing scientists and inventors of the past.

He was pulled into UFO research after seeing them close up, and what ever you show him, he figures out because of his inquisitive, mechanical, scientific mind. These things happened to the right guy because he knows how to explain them well, and how to detail what he saw accurately, being a skilled observer. One of the many millions of UFO eyewitnesses across the world, his story could be anyone's including yours, and thanks to his scientific knowledge he offers us a unique perspective on this phenomenon. Because of his in depth knowledge of the Bible and supporting ancient scriptures, he gives a good insight in a biblical perspective on the implications of this subject. His perspective is a new one. He says that it is the citizens, the individual eyewitness researchers, who are the true experts on the subject of UFOs. He has an in depth knowledge of suppressed technology and high-tech inventions of the past, and can explain the mechanics of just about all UFO experiences in relation to Earth technology.

One of his friends is a former government physicist who worked on top-secret projects. Creative intelligence and insight are the ultimate qualifications to this author, combined with humbleness towards the creator. After speaking to so-called educated people who have made fools of themselves, and also speaking to so-called uneducated people who are brilliant, this author has concluded that those who seek the truth and submit to the spirit of truth, are the ones who are given true insight. He has tied together this research with much insight, and the help of a fantastic memory regarding previous cases and research information.

Sources

The following books have been a great help to me in my research and are recommended reading.

- Manmade UFOs 1944 to 1994 -50 years of suppression by Renato Vesco and David Hatcher Childress
- Genesis by W.A. Harbinson
- Genetic Armageddon by Stephen Quayle
- Genesis 6: Giants by Stephen Quayle
- German Secret Weapons of WW Two by Rudolph Lusar
- Intercept but Don't Shoot, by Renato Vesco
- UFO's,Nazi Secret Weapon? By Mattern&Feidrich
- Space Aliens from the Pentagon by William R. Lyne
- Missing Time by Bud Hopkins
- Anti -Gravity & the World Grid by David Hatcher Childress
- Casebook on Alternative 3 by Jim Keith
- Vimana Aircraft of Ancient India &Atlantis by David Hatcher Childress
- Anti-Gravity The Dream Made Reality by John Thomas Jr.
- The Fantastic Inventions of Nikola Tesla, edited by David Hatcher Childress
- The Truth behind the Davinci Code by Richard Abanes

- The Davinci Deception by Irwin W. Lutzer
- The Chester Beatty Papyri ... (7 volumes and numerous articles) by Sir Frederic Kenyon
- The Holy Bible... contains the only books of 100% accurate prophecy inspired by Holy Spirit (66 books)

Internet articles referred to...

- The Brave New World of Scalar Electromagnetics by Col. Tom Beardon on Jeff Rense's web site
- The True Story of Tesla and his murder by Webster Tarpley (also on Jeff Rense's web site)
- A Solar Radiation Powered Gravity Field Generator by J. F. Sutton available on the Hutton Commentaries, Inc.
- Strange Craft article on Coast to Coast AM